DR STUART CLARK is the author of the award-winning book *The Sun Kings: The Unexpected Tragedy of Richard Carrington and the Tale of How Modern Astronomy Began*, and the critically acclaimed *The Sky's Dark Labyrinth* trilogy. A former editor of the United Kingdom's bestselling astronomy magazine, *Astronomy Now*, and a Visiting Fellow at the University of Hertfordshire, he writes the *Guardian*'s astonomy blog, *Across the Universe*, which attracts around three-million readers a year. His previous books include *The Unknown Universe*.

THE SEARCH FOR EARTH'S TWIN

THE EXTRAORDINARY, CUTTING-EDGE STORY OF THE SEARCH FOR A DISTANT PLANET LIKE OUR OWN

STUART CLARK

Quercus

First published in Great Britain in 2016 by Quercus.
This paperback edition published in 2017 by

Quercus Editions Ltd
Carmelite House
50 Victoria Embankment
London EC4Y 0DZ

An Hachette UK company

A CIP catalogue record for this book is available
from the British Library

PB ISBN 978 1 84866 585 9
EBOOK ISBN 978 1 78429 035 1

10 9 8 7 6 5 4 3 2 1

Typeset by CC Book Production

Printed and bound in Great Britain by Clays Ltd, St Ives Plc

Contents

ONE

If nature can do it, so can you

It was clear to the young astronomer that he was having some sort of breakdown. It was 1982 and Geoffrey Marcy was in his late 20s. Ironically, his dream was coming true; he was working with the most qualified professors, at the most prestigious observatories in America, yet he was weighed down by feelings of inadequacy and incompetence.

He was discovering that science wasn't so much collaborative as cut-throat, and he wasn't sure he had the strength to fight. His work was being publicly criticized, and he was in therapy for his emotional state.

Academia had seemed such a natural path for him growing up, and astronomy was the discipline he fell in love with. He had lived his early life in the Los Angeles suburbs of the San Fernando Valley during the 1960s. His mother and father were both well educated with specialisms in anthropology and aerospace engineering. Neither was afraid to speak up for what they believed in. His mother was an active supporter of the civil rights movement for minorities. His father would enthuse about the latest aerospace technologies such as supersonic flight or the space shuttle.

When Marcy was a boy, they bought him a small second-hand telescope. At night, he would climb out of his bedroom window onto the patio roof, carrying the present, so that he could observe the stars. At school he struggled to achieve in anything other than the physical sciences, so the correct path for him was clear. He followed it right into the physics faculty at the University of California, Los Angeles (UCLA), where he found the work challenging but satisfying, and so continued on to UC Santa Cruz as a graduate student. There he was taught the art of observing stars with large telescopes and worked towards a Ph.D. studying the magnetic field of stars like the Sun. He was supervised in these efforts by Steven Vogt, who had tried the self-same project for his own Ph.D. a decade before – and failed because the technology was not up to the job. Vogt had therefore dedicated himself to developing better instruments and, in the intervening years, succeeded to the point where he thought it was worth another go.

Under Vogt's guidance, Marcy became a skilled observer and succeeded in detecting these distant magnetic fields. His work earned him a Carnegie Fellowship at the Mount Wilson and Las Campanas Observatories in Pasadena. But it was here that things started to go wrong.

His Ph.D. work, which had originally seemed so successful, was coming under sustained criticism by a respected astronomer from Harvard. It triggered feelings of inadequacy in Marcy. He couldn't seem to find his feet, and his confidence began to tumble; everyone, it seemed, was smarter than he was, and he began thinking that the game was up. Any day he would be found out as a fraud who had been lucky to get as far as he had. With the self-doubt lodged firmly in his mind, his mental health deteriorated. He thought his career was over.

Desperate for a meaningful, yet less competitive way of life, he was lingering in the shower one morning, talking himself into going to work. That's when it hit him as searingly as a shaft of the dawn sunlight. Why not subvert the observational techniques he had learnt towards answering an almost forgotten question in astrophysics: how many planets like Earth were there in the Galaxy?

Reason alone suggested such planets should be manifest. It was almost inconceivable that our Sun should be unique in this respect but proving it was a different matter entirely. The technological requirements were so exacting that Marcy knew of no one who was pursuing the goal. If anything, it was a discredited science. Since the 19th century there had been a number of astronomers who had claimed to have discovered planets but all had turned out to be embarrassingly wrong. Yet to Marcy, this was a freeing concept. His career, he thought, was over anyway. So why not spend what remained of it in pursuit of something that had little or no hope of success but in which he could at least believe.

Detecting planets is not yet as simple as pointing a telescope in their direction and taking an image. It relies on the Doppler effect. This phenomenon is a staple of school physics classes, both because it is fundamental to the way we perceive the world and because it is easily demonstrable with sound. Think of the whine of a siren as an emergency vehicle passes you in the street. On approach, the sound is squeezed, giving it a higher pitch. Once the vehicle has passed the opposite happens and the sound waves are stretched to lower pitch. Listening, we hear this as a continuous change when in reality the siren has not changed its pitch at all. The sound is modified because of the speed and direction of the vehicle.

The effect owes its name to Christian Andreas Doppler, a 19th-century physicist who is persistently mis-named Johann Christian Doppler because of a mistake in the writings of influential German astronomer Julius Scheiner.

Doppler was born on 29 November 1803 in Salzburg, Austria. He was never in the best of health and this presented his family with a problem. His father ran a successful stonemasonry and in the normal run of things would have expected his son to continue in the business. Christian's physical frailty ruled out such demanding physical work, so they decided that academic education was the best route. Using some of the family's accumulated wealth, they sent the boy away to secondary school in nearby Linz, and then to the Vienna Polytechnic Institute, where he showed a flair for mathematics.

By 1829, Doppler had successfully studied mathematics, mechanics and astronomy and was looking for a permanent academic job. It was not easy; he was passed over time and again. The years slipped by and he found himself working as an accountant in a cotton-spinning factory. Growing despondent, he applied to emigrate to America. In preparation, he began to sell the possessions he had accumulated. Yet, at the last moment, he was offered a post teaching mathematics in Prague. He took it and over the next few years climbed the career ladder to end up as professor in practical geometry and elementary mathematics at the Prague Polytechnic. It was here in 1842 that he made his great breakthrough.

There was a growing interest in understanding the nature of light. It was widely thought that light was a wave motion and that colour was a reflection of wavelength. Doppler's genius was to recognize that the emission of the wave would take some time, so if the emitting object moved during that process, the light's wavelength would be altered. It would either be squashed or stretched depending upon whether the light source was moving towards or away from the observer. The same would happen if the observer were moving because it would take time to absorb the wave. Any change in the wavelength of the light would translate into a change of colour.

While physicists were interested in light, astronomers were fascinated by 'double stars'. The general wisdom of the time was that most stars emitted more or less yellow light but there were a number of double stars in which the light components were strikingly different colours. Alberto in the constellation of Cygnus was seen through telescopes to be two stars closely separated, one yellow and the other blue. The same is true for Almach in the constellation of Andromeda. There are subtler variations too. Epsilon Lyrae is composed of four separate stars, leading astronomers to call it the double double. Each one is a different shade of blue. Some individual stars are distinctly non-white too. Betelgeuse and Rigel, both in Orion, are red and blue respectively.

Doppler wondered whether motion was responsible for these different colours. For the double stars, orbital motion could be squashing the yellow light, turning it blue. In the case of the single stars, their movement through space could be altering the hue. He found that he could derive a formula that related speed to the change in wavelength, and presented it in a lecture given to the Royal Bohemian Society on 25 May 1842. It was accompanied by a paper titled 'Über das farbige Licht der Doppelsterne und einiger anderer Gestirne des Himmels' ('On the coloured light of the double stars and some other stars of heaven').

He likened the phenomenon to ships in choppy waters. When navigating into the waves, a ship would encounter the peaks more rapidly than when motoring along with the current. In the first case, this increase in frequency translates into an apparent shortening of the wavelength (the distance between adjacent peaks). In the latter, the opposite situation occurs.

Doppler stated that, if it was correct, his hypothesis could provide astronomers with a means of calculating the speed at which stars were moving. The question was how to prove it. You could not go to the stars and see if they really were moving. Nor could you control light accurately enough in a laboratory; the apparatus of the day was simply not up to the task. Fortunately, Doppler's formula was completely general. His proposed effect would take place on anything that was composed of waves: sound, for example. But it wasn't Doppler who translated this into an experimental test; it was a Dutch polymath and a group of musicians.

Christophorus Henricus Diedericus Buys Ballot was born in Kloetinge, The Netherlands. He studied for a degree at the University of Utrecht and remained there for his whole career. He studied geology and mineralogy but is mostly know for his contributions to chemistry and meteorology.

In 1843, he rounded up a group of trumpeters, all of them with perfect pitch, and took them to the town of Maarssen, a stop on the Utrecht–Amsterdam railway line. There, the authorities placed a locomotive at his disposal, and he loaded some of the musicians onto an open truck and had others stand at the side of the track. The locomotive was to drive repeatedly to and fro, with the musicians on the truck blowing their trumpets and those on the side listening and commenting on the pitch. The trouble was it was chilly in December and a snowstorm soon convinced them all to return home.

The experiment finally took place, with more people to listen and play, in June 1845. This time, they were beset by the summer heat but the experiment went ahead. Up and down went the train, altering its speed between the runs. Sometimes the trumpeters on the train would play and those at the trackside would listen, other times vice versa. At the end of it all, there was no doubt. Doppler was right, waves were modified by movement between the source and observer. This confirmation was, quite literally, accompanied by a fanfare.

The Doppler effect is now ubiquitously used to measure movement in all scientific disciplines. Detectors for monitoring a patient's heartbeat or an unborn child measure blood flow by bouncing ultrasound waves off the moving blood cells. Radar detection, which tells us the speed and direction of an object, is based on the Doppler effect, as are police speed guns. Musically, the effect was pressed into action by inventor Donald Leslie, who developed a rotating loud speaker in the 1940s to change a pure sound into a more ethereal one. The Leslie speaker became a staple for Hammond organ players throughout the 1950s and also gave George Harrison his distinctive guitar sound on the Beatle's 'Lucy in the Sky with Diamonds'.

But for the effect to be used with the stars, as Doppler had originally hoped, some way of knowing the original wavelength at which the stars were emitting light had to be found. This too was discovered in the 19th century.

*

The path started with English physician William Hyde Wollaston, whose career only really took off after he had retired from the profession and reinvented himself as a chemist. One of 17 children, Wollaston had graduated from the University of Cambridge and practised medicine, but the suffering of his patients weighed so heavily on him that he began looking for a way out. Salvation came when he received a large gift of money from one of his brothers, allowing him to give up medicine and divert his energy to chemistry, his true love.

At the turn of the 19th century, having moved to London, he established a company that traded in chemical products, and he became an influential member of the Royal Society, which had been founded in November 1660 to pursue the investigation of nature. He stood before the assembled fellows on 24 June 1802 and related a most curious discovery he had made while investigating the light-bending properties of various substances. Taking his cue from some of Isaac Newton's work that had been described to the society back in 1672, Wollaston had blacked out a room, except for a tiny pinpoint of light at the window. Whereas Newton had passed the shaft of sunlight through a prism and shone the result onto the far wall, Wollaston stood some 10 or 12 feet away and observed the light through a prism of flint glass that he held to his eye.

In Newton's classic work on colours, the Cambridge don described how a shaft of white light was split by the prism into red, orange, yellow, green, blue, indigo, violet, each colour merging into the other. He had called this pattern a spectrum, a Latin word that means image or apparition, even spectre. Wollaston, in contrast, described how he had seen just four colours in the spectrum: red, yellowish-green, blue and violet. Most extraordinarily however, Wollaston told of how these colours appeared to be divided by black vertical lines.[1] Intrigued, he had repeated the experiment with candlelight, and discovered that the pattern changed from black lines to bright lines of colour. He concluded his paper by confessing that he could not explain the phenomenon. No one could; it was entirely unanticipated and the matter rested for just over a decade until the German physicist Joseph von Fraunhofer took up the challenge.

Born on 6 March 1787, Fraunhofer was orphaned at 11 and became an apprentice to a glassmaker in Straubing, Bavaria, who worked him like a slave. One day, disaster struck when the glassworks collapsed. Fraunhofer was pulled from the rubble under the gaze of Maximilian IV Joseph, the Prince Elector of Bavaria, and taken into his royal patronage. In 1806, he moved to a glassworks in the secluded Benediktbeuern Abbey, a former Benedictine monastery some 65 kilometres south of Munich. Following the secularization of Bavaria in 1803, it had become a great industrial research and development enterprise with the goal of making the finest glass in the world.

Fraunhofer thrived, working in the toxic environment of the furnaces to develop new recipes for different types of glass. A key property for glass was its dispersive power, the measure of how widely it could spread the colours of a spectrum. In 1814, Fraunhofer invented a device called a spectroscope to measure this property in samples of glass.

It was a tabletop device that allowed a prism to be fixed into place and observed through a small telescope, which magnified the resulting spectrum. Using the Sun as his source of light, Fraunhofer rediscovered Wollaston's dark lines. As he developed the spectroscopes further, he found that increasing magnification revealed more dark lines in the Sun's spectrum. In all he counted 574 of them, and they are still known today as the Fraunhofer lines. In time, astronomers would come to realize that these spectral lines are essential for measuring the movement of stars, and inferring the existence of planets around them, but at the time their origin was a complete mystery.

Fraunhofer himself never found out what the spectral lines were. Like so many of his profession, he died young, perhaps as a result of working in the poisonous fumes of the glass furnaces. He died aged 39 on 7 June 1826.

The next leap in understanding did not come until the middle of the century, more than 300 kilometres away in Heidelberg.

German scientists Robert Bunsen and Gustav Kirchhoff were working on techniques to prepare and ensure the purity of chemical samples. A widely used technique among chemists at the time was the flame test. This involved dropping a small sample of a substance into a flame and seeing what colour it flared. For example, table salt flares bright yellow because of the sodium it contains; calcium burns brick red. The test wasn't definitive because many elements cause similar colours – aluminium and cobalt both give off a silver-white light – so the question was how to progress the analysis to reliably separate these identifications.

Bunsen and Kirchhoff took as their starting point work that had taken place in England in the year after Fraunhofer's death. John Herschel and William Fox Talbot had passed light from various flame tests through a prism and seen that instead of a continuous rainbow of colours, each chemical element gave off a unique pattern of coloured lines when burnt. They wrote, 'A glance at the prismatic spectrum of a flame may show it to contain substances which it would otherwise require a laborious chemical analysis to detect.'

In the case of lithium and strontium, both of which burnt with a red flame, when this light was passed through a prism, each element's light resolved into a different pattern of red lines. It was as distinctive as a fingerprint. Bunsen and Kirchhoff extended this investigation and showed how the bright lines of the flame tests were related to the dark Fraunhofer lines.

Fraunhofer had labeled some of the more prominent dark lines with letters. One of these, the D line, cuts through the orange part of the spectrum. Because sodium burns with a bright orange flame, some wondered whether the D line was somehow also linked to sodium. Instead of emitting the orange light, it was somehow absorbing it. There was also a strong similarity between potassium's bright red lines and a group of dark lines clustered around what Fraunhofer had called the A line. But the question was how to prove this, and why were some lines bright while others were dark.

The breakthrough came in 1859. Kirchhoff directed a beam of light from burning lime through a prism. The lime sample gave off a brilliant white light (this is why limelight was used to illuminate theatre stages). When passed through the prism, it produced a continuous spectrum of colours on a screen at the other end of the lab. To perform his experiment, Kirchhoff positioned a bunsen burner between the light source and the prism so that the beam of limelight passed through the flame before it hit the prism. The continuous spectrum remained. Next, he peppered the flame with sodium. As expected, it changed colour to become the chemical's characteristic orange light but the revelation was what happened on the screen. When the sodium was burning in the flame, Fraunhofer's black D line appeared in the spectrum on the screen. It showed that the sodium was absorbing that specific orange wavelength from the limelight and blazing it around the lab in the form of the yellow flame.

So the dark lines signalled when light was being absorbed by specific elements, and bright lines showed when it was being emitted. Physicists started to refer to the unbroken rainbow of colours as a continuous spectrum. When there were dark absorption lines present, they called it an absorption spectrum because it showed that the light was passing through an intervening cloud of absorbing chemicals. A pattern of only bright lines, such as given by passing a flame test's light through a prism, was termed an emission spectrum.

Kirchhoff tried the experiment next on sunlight rather than limelight. The familiar pattern of Fraunhofer lines naturally present in sunshine appeared on the screen. This time he chose lithium to drop into the flame because, unlike sodium, it was a chemical that did not appear to have a corresponding Fraunhofer line. Dusting the flame, he watched a dark line appear in the red part of the spectrum, nestled between the naturally occurring absorption lines. At a stroke, Kirchhoff proved two things: there had to be sodium in the Sun because of the presence of Fraunhofer's D line, but lithium was missing because of the lack of a lithium line.

The spectral lines, whether absorption or emission, were unique to each element. They were indeed a chemical fingerprint, which meant that by studying the light from celestial objects astronomers could work out the distant orb's composition. It was a watershed for the science because it gave the astronomers a way to investigate the very nature of the celestial objects rather than just chart their positions.

More than this, the spectral lines gave physicists and astronomers a way to measure the Doppler effect on starlight. This was because the spectral lines were fixed points of reference that occurred only at specific wavelengths. If the stars were moving to or from Earth, the lines would be shifted from those wavelengths.

Astronomers began building spectroscopes for their telescopes, firstly to analyse the stars chemically and then, as the instruments became more accurate, to measure the Doppler effect on them. The obvious first targets were the binary stars, as Christian Doppler had first suggested.

In an orbiting pair of stars, one would always be moving towards Earth, while the other was moving away. This meant that the Doppler effect would always be blue-shifting (squashing) the light of one, while red-shifting (stretching) the light of the other. Indeed, astronomers began to clearly see Doppler shifts in the binary stars but the shifts were too small to account for the changes in the stellar colours. Nevertheless, the technique opened up a whole new way of investigating stars and, by the first half of the 20th century, a key goal of many of the major observatories around the world was to conduct surveys of the sky, measuring the movement of the stars towards and away from us using the Doppler effect.

By the middle of the 20th century, about the same time that the Doppler-inspired Leslie speakers were taking the popular music world by storm, astronomers were turning away from the technique because most surveys had been completed.

German-born astronomer Otto Struve thought that this was a lost opportunity and penned a two-page paper to describe a bold new use for the Doppler effect in stars: it could be used as a method for detecting planets.

His paper is notable because it has the overtones of an impassioned plea. He writes, 'One of the burning questions of astronomy deals with the frequency of planet-like bodies in the galaxy which belong to stars other than the Sun.'[2]

Struve believed that extraterrestrial life was probably widespread in the Galaxy and had formed this opinion because of his studies into the rotation of stars. To measure the rotation of a star, it is necessary to take a spectrum and use the Doppler effect. The star itself cannot be resolved by the telescope because it is very far away, so the light from the entire surface of the star arrives together in a single beam. As the star turns, one hemisphere will be rotating towards the observer, while the other will be rotating away. So the side coming towards us will squash the light and the side moving away will stretch it. This will broaden the spectral lines by making them cover a wider range of wavelengths than normal. Struve had been measuring this broadening and using it to calculate the rotation speed of the star.

He found that most stars rotate slowly, rather like the Sun, which takes almost a month to spin once. Such sluggish behaviour runs against expectations because, if stars are condensed gases from a widely spread-out cloud, then they should be rotating extremely quickly. It is the result of something known as the conservation of angular momentum.

In the case of the Sun, astronomers had reconciled its slow rotation with the fact that the Sun is surrounded by planets. Indeed, although the mass of the Sun makes up more than 90 per cent of the mass of the entire solar system, the proportion is reversed when it comes to the angular momentum. The planets in their orbits carry most of the rotational energy. Struve thought that the same must be true for other slowly rotating stars. So finding planets around them was a key test of his hypothesis.

Although planets are far too small and dim to be seen directly by the telescopes of his time, Struve pointed out that, while the Sun's overwhelming gravity pulls the planets into orbit, the planets' gravity has a reciprocal, though much lesser, effect on the Sun. Jupiter is forced to move at a speed of 13,000 metres per second around an orbit that takes 11.86 years to complete. In return, Jupiter pulls the Sun into a pirouette that similarly lasts for about a dozen years but moves only about 13 metres per second, roughly the sprinting speed of a world-class athlete. Although fast by human standards, this is painfully slow by celestial measures.

Struve pointed out that the star's motion in response to the planet's gravity would produce a changing Doppler shift on the star's light. During the pirouette, the star would sometimes be coming towards us, other times moving away, and this would result in the spectral lines moving from one side to the other of the spectrum. Although this movement was far below the detection threshold of the spectroscopes of the time, there was no reason why Jupiter-sized (and larger) planets might not be much closer to their parent stars. This would generate much larger orbital velocities, which it might just be possible to measure.

Struve even calculated that there was nothing to stop a planet being extremely close to its parent star and completing a fast and furious orbit in just a day. How prescient these thoughts were in light of the discoveries that were to come, but at the time his suggestions were largely ignored. Stellar astrophysics in the 1950s was largely concerned with calculating the details of how stars generated their energy by converting one chemical element into another; no one went looking for planets.

That changed when Marcy figured he had nothing to lose.

Working at Mount Wilson gave Marcy access to one of the most historic telescopes in the world. Constructed in 1917, the 100-inch Hooker telescope wears its age with pride. Its riveted metal frame gives it an almost Art Deco feel. Aesthetics aside, it has a special place in astronomy because American astronomer Edwin Hubble used it in the 1920s to show firstly that galaxies were distant collections of billions of stars, and secondly that the universe was in a constant state of expansion.

Mount Wilson is a 1,740-metre peak in the San Gabriel Mountains. A century ago, it was surrounded by nothing but darkness. Now, the urban sprawl of Pasadena and Los Angeles has reached the mountains, and light pollution has severely restricted the kind of work that the telescope can perform. The faint galaxies that Hubble had made Mount Wilson's stock in trade are lost to sight, and only bright stars can now be reliably observed.

But these were exactly the kind of stars Marcy was interested in. He began to look at the movement of their spectral lines to gauge how precisely he could measure a star's motion from their Doppler shift. What he found was initially not encouraging.

Although he knew he was working with the most precise spectroscope in the world, it only appeared capable of detecting Doppler shifts that corresponded to a velocity of 300 metres per second and above. This was just not enough. Jupiter moved our Sun at just 13 metres per second.

After two years of struggling at the telescope, Marcy finally realized the problem. There was nothing wrong with the instrument as such; it was capable of far better precision. The problem was the Earth's atmosphere. When you look up into a clear sky, the stars appear to twinkle. This effect is caused by the starlight bobbing through the turbulent layers of Earth's atmosphere. To the naked eye, it may look attractive but it foxes astronomers because the twinkling stars appear to dart around when magnified.

Marcy discovered that this was the major limiting factor in the spectroscope's detection because it blurred the resulting absorption lines, making them more difficult to measure accurately. It was a bittersweet revelation because it meant that simply building a better spectrometer wasn't necessarily the route to better observations. Instead, what you really needed to do was remove Earth's atmosphere so that the star would stop swimming around.

All in all, it looked to Marcy like the naysayers could have a point after all. But to someone who already thinks he is a failure, such an inconvenient truth was not necessarily a reason to give up. Also, in the back of his mind was a phrase that a professor from his undergraduate years had used on him. At the time, Marcy had been despairing over failing to complete a calculation involving heat flow through an interstellar gas cloud but the professor had refused to let him admit defeat.

His words were again echoing through Marcy's mind: 'If nature can do it, so can you.'

But how?

Marcy's fellowship at Mount Wilson was coming to an end and he knew that if he were to continue the pursuit of planets, it would mean making a sacrifice. Such a project was just too speculative for the major establishments who employed him. They demanded publishable results on an almost monthly basis. A planet search would involve an extended period of technological development during which there would be little to publish. So he took a teaching job in the less pressured environment of San Francisco State University. There, working under the cover of his teaching commitments, he quietly began the programme that would help change our perception of the universe forever.

It was slow going at first; he simply could not see a way of combating the blurring effects of the atmosphere. Then two great things happened. The first was that he went to a lecture given by a visiting astronomer named Bruce Campbell.

Campbell was working with another astronomer named Gordon Walker. Both were at the University of British Columbia in Canada and had developed a new technique that involved passing starlight through a transparent container of gas just before it hit the spectrometer and was recorded. The gas they used was not for the faint-hearted. It was hydrogen fluoride, a highly corrosive substance that would transform into hydrofluoric acid upon contact with moisture, including that found naturally in human skin.

Despite the dangers, they used it because it was a compound that was not naturally found in stars, so it superimposed a unique sequence of spectral lines on the starlight. These could be compared with the stellar lines, allowing wavelengths to be measured very accurately. Campbell and Walker then took many short exposures rather than fewer lengthy ones. These short exposures 'froze' the spectrum before the turbulent atmosphere had time to blur the lines, allowing the precise wavelength to be pinpointed more easily. Combining the results from all the short exposures allowed them to measure stellar motions to an accuracy of 10 metres per second.

This was a number to make astronomers sit up and take notice because it was about the calculated velocity at which Jupiter's gravity pulls the Sun. It wasn't the end of the story, because to make a gold-standard detection, scientists talk about 5-sigma discoveries, where sigma (σ) stands for the signal-to-noise ratio. The precision of the instrument is effectively the noise level. So a precision of 10 metres per second means that this is the magnitude of the errors, or noise, that you can expect the instrument to give. To be absolutely certain that you are not being fooled by random noise, the signal has to be five times larger than the noise, in this case, 50 metres per second. Although you can be reasonably sure of a discovery at the 2–3σ level, you are not watertight until you reach 5σ. So it was impossible to detect Jupiter-like planets with Campbell's spectrometer – that would require a precision of 2 metres per second – but the instrument was clearly nudging at the limit, placing the detection of giant planets on the technological horizon.

Marcy was dumbstruck. This was the breakthrough he had been searching for. He had to find some way to replicate what he was doing and so began talking to other astronomers about the technique. In the course of these conversations, he discovered that the solar physicists used a similar technique with iodine, a far less hazardous gas.

The second great thing that happened was that he met the

student Paul Butler, an undergraduate studying both physics and chemistry. Butler's interests and skills were perfectly fitted to Marcy's ambition, and they began to develop an iodine cell to sit in front of a spectrometer and provide the reference wavelengths. Once they had the hardware, however, the next hurdle was how to test it. Every sensible astronomer 'knew' that searching for planets was beyond the grip of a spectroscope even of this precision, so no allocation committee would grant any time on a telescope with that as the aim. If they were going to get any time under the night sky, they needed a different angle.

As luck would have it, astronomy at the time was alive with the idea of 'dark matter'. This mysterious stuff was thought to be spread widely across space because almost everywhere that astronomers measured movement they detected too much. Only the planets in our solar system moved in a way that could be completely understood. This remains the case today. Stars and galaxies orbit faster than the visible matter suggests is possible. The most obvious way to speed things up is to add more matter, which generates more gravity, pulling things into faster orbits. Since the matter could not be seen, it was reasonable to suppose that it was not emitting much light.

There are two possibilities for this dark matter: either it is a cosmic ocean of exotic subatomic particles or it consists of extremely dim celestial objects such as small stars that have failed to ignite or the mysterious stellar corpses called black holes. The failed stars are called brown dwarfs. They have condensed from the gas clouds of interstellar space just like other stars but they have not managed to accumulate enough mass to spark nuclear reactions at their heart. Without these reactions, they cannot generate energy and so cannot shine. A brown dwarf sits in space as nothing more than a big sphere of gas, rather like a bigger version of a gas giant planet. This fascinates astronomers: brown dwarfs form like stars yet end up like planets, putting them right on the dividing line between the two classes of celestial objects.

Because they do not radiate visible light, however, brown dwarfs are very difficult to see. So astronomers were not sure how many of them there are. The only guide they had was that there are more smaller stars than larger ones. If brown dwarfs followed this trend then there could be an overwhelming number of them, yet in the 1980s astronomers had not glimpsed a single one for sure. Being larger than Jupiter by some 10–50 times, they would generate a much larger wobble in their parent star than a planet. As they had a suspected velocity of many dozens of metres per second, Marcy and Butler's new device stood a chance of seeing them.

So Marcy wrote the science case saying that he hoped to detect substellar objects, a euphemism for brown dwarfs. When the responses came back, however, they were not all he might have hoped for. He had been granted time but it was the least possible time you could imagine: one night every month or so at the University of California's Lick Observatory, and right at the time of the full moon. Frankly, it was time that nobody else wanted and he suspected that he was given the time just to avoid the telescope standing unused. But it was a start.

By now it was 1987, Butler had received bachelor's degrees in physics and chemistry from San Francisco and was working towards a master's degree in physics, but he was soon going to have to leave because the university did not offer doctorates in the subject. Having started to take data, Butler began writing computer code to analyse the results. It was a long-winded process and took many years of trial and error. Both he and Marcy were perfectionists and, besides, they did not think that they had any reason to rush.

Contrary to Struve's open mind of a century before, the theoreticians were very clear that a planet like Jupiter could only be built far from its parent star, where the gas was cold enough to condense. So any Jupiter-sized planets would be orbiting their stars at 3 to 5 times further away than Earth is from the Sun. That meant that it would take a giant planet 10–15 years to go round its star. Since the astronomers would need more than a full orbit to be sure of a detection, and the spectroscope wasn't yet of the accuracy needed to see the planets in any case, they assumed there was no hurry. So they spent their time collecting data and perfecting their analysis software instead of searching the data for signals. That was a mistake.

What they didn't know was that across the Atlantic a pair of Swiss astronomers were puzzling over some data of their own, and coming to a startling conclusion.

TWO

Impossible planets

Michel Mayor was in his early 50s at the beginning of the 1990s. With beard, wire-rimmed glasses and a domed pate, he looked like the kind of professor you imagine buried behind a pile of books, or indeed squinting into the eyepiece of a telescope. Rubbing against this stereotype of a detached academic was his endearingly boyish grin.

Mayor was born in Lausanne, Switzerland, in 1942 and after receiving a degree in physics from his hometown university in 1966, he moved to the University of Geneva. Throughout much of the 1980s, Mayor had been searching for brown dwarfs, the failed stars that had gripped astronomers' attention. In 1989, he was one of a team of scientists who discovered that the star HD 114762 was wobbling. Back-calculating what kind of object could be responsible for moving the star like this, the team discovered that there was a brown dwarf of between 11 and 145 times the mass of Jupiter in an elliptical 84-day orbit around the star.

This helped spur Mayor's ambition to develop a new spectroscope that could reach even better precision, and thus reveal smaller brown dwarfs, paving the way to planets. Astronomical technology was progressing quickly in the late

1980s and early 1990s. The pixels in electronic cameras called charge-coupled devices (CCDs) were getting smaller and smaller, meaning that they could record wavelength information more accurately. At the same time, computers were getting faster and so were able to handle the larger amounts of data that CCDs were producing.

In the early 1990s, Mayor took on a Ph.D. student to help him develop the next generation of precision spectroscope. Didier Queloz was born in the same year that Mayor graduated. He was also a physicist turned astronomer. There are few photos that do not show him smiling, with his long face and a dense shock of black hair. Even now that he is a distinguished professor of physics at the University of Cambridge, UK, and approaching his fiftieth birthday, he still radiates all the enthusiasm of youth, finding it difficult to sit still as he relates the story of his extraordinary discovery.

Mayor and Queloz set to work on building a state-of-the-art spectroscope that they named ELODIE. From the very beginning, it was destined for the Observatoire de Haute-Provence's 1.93-metre-diameter telescope. In 1993, the astronomers were ready for 'first light', the moment when starlight is focused onto the instrument for the very first time. ELODIE performed better than its design specification, reaching a precision of about 13 metres per second. For Queloz, the figure brought with it a moment of revelation: the detection of planets was now a possibility. A true Jupiter was still outside the envelope of detectability but, like Otto Struve decades before, Queloz wondered about a giant planet that was a little closer to its star or a little bigger than Jupiter. A planet such as that could become detectable, especially with repeated observations, because the gradual accumulation of results could help achieve even greater accuracy with the same instrument.

The trouble was that the observation campaign that Mayor had in mind relied on the observation of a large number of stars, instead of dwelling on just a few for a long time. This was because the suspected brown dwarfs would be very obvious to ELODIE, and so would show up quickly, allowing a large number of stars to be surveyed. Searching for planets around a few arbitrarily chosen stars was a much riskier way to use their observing time. There was no guarantee of success and Mayor worried that it was too much of a risk for a student working towards a Ph.D.. Both astronomers were aware of Marcy and Butler's programme, and of Campbell and Walker. Neither of those teams was finding planets, which only confirmed to Mayor the need for caution. Yet Queloz could not get the idea out of his head. He implored his supervisor to let him take the risk.

In 1994, Mayor was offered a sabbatical in Hawaii. He agreed to let his young colleague start a planet search with ELODIE during his time away. When he returned to Geneva, they would review progress and, if needed, return the research to brown dwarfs in order to ensure that Queloz could achieve his Ph.D.. Left on his own, Queloz set to work with a volcanic sense of energy and enthusiasm.

Not only was he working at night with the spectroscope to collect the data, he was also writing thousands of lines of computer code that turned the electronic readings from ELODIE into stellar velocities. It was painstaking and lonely work. He chose his targets carefully, so as not to overlap with Marcy and Butler, who had recently published a list of their target stars, and in September 1994 he turned the telescope towards 51 Pegasi for the first time. This is a star similar to our Sun that lies about 51 light years away. It is just visible to the naked eye under excellent viewing conditions and was first catalogued in 1712 from the Royal Greenwich Observatory, London, by John Flamsteed, the first Astronomer Royal.

Queloz decided to concentrate on 51 Peg for a little while because it was bright to the telescope, so could give excellent data, and he thought this would be a great way to understand more about the behaviour of the spectroscope. It was a thoroughly dispiriting endeavour, however, because after collecting just a few data points, he saw that the star appeared to be unstable. Its velocity was changing by large amounts over just a few days. He thought he had perhaps made an error in his coding. The idea was accompanied by a sense of panic. This was his Ph.D. at stake. He remembered that Marcy and Butler had been working for the best part of a decade on developing their analysis software, yet he had put his together in about a year.

Doubts crowded his mind and he began to consider where the error could have crept in. There were so many possibilities that it was overwhelming. Then he thought of a way to double-check the problem. He should stop second-guessing and just press on with his observations, looking at different stars to see if the same pattern of variation was repeated. By the end of the year, he had collected enough data from enough different stars to know that it probably wasn't a glitch in his software. Only 51 Pegasi showed this large variation in velocity. He was at the observatory collecting yet more data towards the end of the year when it dawned on him that since he had eliminated technological faults the signal had to be real, and that meant 51 Pegasi was being pulled around by an unseen companion. He could not say what the mystery object was because he had not yet written the software needed to analyse the motion and deduce the orbit and mass of the object. Then the weather rolled in.

For two days the observatory was wreathed in electrical storms. Absolutely no observations were possible. Queloz headed for the library. Underneath the crack of thunder and the strobe-like illumination of the lightning coming through the windows, he read up on techniques and wrote a computer routine to extract the orbital characteristics of the body from the data.

When he applied it to his data, the computer solution revealed a planet about the mass of Jupiter. It was an eerie moment. He told no one because he was too afraid of being wrong. Instead, he came back to the observatory in January to collect more data. Using the orbit that the computer had calculated, he predicted what he should see in January if his solution was correct. However, the January readings differed from his prediction. He had not yet homed in on the right solution. It took him until March 1995 to collect enough measurements and find an orbit that made a successful prediction and that he could be statistically confident about. But even then, it looked all wrong.

Instead of a decade to complete an orbit, the Jupiter-sized planet that Queloz had found circled its parent star in just 4.23 days. This was extraordinary, because the speed of an orbit is more to do with how close the planet is to its star than the speed at which it is travelling. So to complete an orbit in just 4.23 days was astonishing. In our own solar system, the innermost planet Mercury orbits the Sun in 88 days – positively leisurely compared with whatever this thing was around 51 Peg. It smacked of some sort of error. Yet Queloz had now observed other stars quite extensively and found no hint of a variation in those. So he was confident that the signal was real.

Mayor was still in Hawaii at the time, so Queloz faxed him the graph of the motion and a simple message saying, 'I think I've found a planet.' His supervisor faxed back with an equally uncomplicated response, 'Well, why not?' He went on to say that he would be back in Geneva in a month and they would look at the data together to be sure.

Once they were convinced it was not a software error, they started to imagine other possible interpretations. They wondered whether the signal could be pulsations of the star, or cooler regions on the surface called starspots. They analysed their signals looking for these characteristics, and nothing fitted the data as precisely as the planet solution. They were as sure as they could possibly be and set everything down in a paper that planned to submit to *Nature*, the world's most prestigious scientific journal. Then they put the paper to one side for a month until they could re-observe the star one more time. They agreed that they would junk the paper if they did not get exactly what they were expecting. They took their data, analysed the results and found their expected signal. So they submitted the paper.

As per their usual editorial policy, *Nature* sent the paper out in the strictest of confidence to three referees. These arbiters are usually taken from the list of astronomers cited in the paper's references and remain anonymous unless they waive that right. In the case of 51 Peg b, two of the referees were cautiously positive. The third was openly doubtful. He thought that there were errors in the data and recommended that it not be published. Perhaps to prove that this was not a vindictive attack, he waived his anonymity. It was the Canadian astronomer Gordon Walker, whose colleague Bruce Campbell had inspired Geoff Marcy to create the iodine cell.

Mayor and Queloz read his critique and saw his point. There was indeed some possibility of error in the data. So they went back to the telescope once again and re-observed. Adding yet more data did nothing but strengthen their result, and when an amended version of the paper was sent to Walker, he replied within 48 hours saying that he was satisfied that his concerns had been addressed. That was the final hurdle. *Nature* accepted the paper, and Mayor and Queloz prepared to go public. Announcing the discovery of a planet around another star was a risky business. There had been so many false alarms that most astronomers now thought of it as a discredited aspect of the science and were openly suspicious of any claim. Certainly the track record had been anything but good.

William Stephen Jacob was the sixth son of a vicar. He became a cadet in the East India Company's Addiscombe College in 1828, and after completing his military education at Chatham was posted to Bombay (Mumbai). By 1845, he had attained the rank of Captain of the Bombay engineers but his mind was turning to scientific matters. In December 1848 he was appointed Director of the Madras Observatory, where he demonstrated considerable skill in measuring the positions of celestial objects. One star in particular caught his eye.

Known as 70 Ophiuchi, it had been discovered to be a pair of stars in 1779 by the great Hanoverian astronomer William Herschel. In the often stifling heat at Madras, Jacob made his own observations and computed by hand that one star was orbiting the other with a period of 93 years. However, there were discrepancies in the results because he could not find an exact orbital solution.

Puzzled by this, Jacob speculated that the error might be because the law of gravity was different near 70 Ophiuchi. Knowing that this was a radical idea, he went on to suggest 'a simpler mode' to account for the discrepancies: the gravitational action of an unseen companion to the stars. In computing the orbit that such a body would need to follow he found it to be a large ellipse which would take 26 years to traverse.

He finished his paper by saying 'There is then, some positive evidence in favour of the existence of a planetary body in connexion with this system, enough for us to pronounce it highly probable, and certainly good enough for watching the pair closely, to procure, if possible, still stronger evidence.'

By January 1896, maverick American astronomer Thomas Jefferson Jackson See thought he had considerable further proof. He had made his observations at the University of Virginia's Leander McCormick Observatory, which sits atop the wooded slopes of Mount Jefferson in Albemarle County, Virginia. He calculated that an unseen companion with a period of about 36 years had to be present in the system otherwise there was no way to marry up the observed orbit with our understanding of gravity.

He mentioned that two recent searches for the mystery body had come up empty handed, yet was convinced that the perturbation of the visible star was a matter of great interest and a matter for 'the most scrupulous care' until the mystery was resolved.

The irony was that others were thinking this particular quality was exactly what was missing from See's own work.

Forest Ray Moulton had been a graduate student of See's and was familiar with the way he worked. He analysed the three-body system that See was suggesting and discovered that it was impossible for two stars and one planet to hold such a configuration. Furthermore, he knew another former student of See's, called Eric Doolittle, had found an orbit for 70 Ophuichi that dispensed with the need for an unseen companion.

Unfettered by any self-doubt, See had been taking his claims of a planetary detection into the mainstream press, writing for the public that he had a further half a dozen binary star systems that appeared to show evidence for planets.

Upon receipt of Moulton's paper, and unimpressed by See's bluster to sidestep the facts, the *Astronomical Journal* printed a rare public warning that See would be subjected to harsh review in future before anything more of his was printed.

He spent the rest of his career and life somewhere between despair and anger. After a breakdown in 1902, he withdrew completely from the mainstream of astronomy, while still pursuing public acclaim as a great scientist. In 1913, a championing book entitled *Brief Biography and Popular Account of the Unparalleled Discoveries of T. J. J. See* was published. Most thought the book had been written by See himself.

The damage had been done. The stigma of astronomers claiming to have detected planets around other stars was born. Other erroneous claims only nurtured this pervasive opinion.

Barnard's Star is one of the closest stars to the Sun. Lying just six light years away, it was named after the American astronomer E.E. Barnard. In the same year that See was publishing his exercise in self-aggrandizement, Barnard was measuring the movement of the star that now bears his name. He found that it was travelling across the sky extremely quickly by stellar standards. This indicated that it must be nearby and made the diminutive red star something of a focus for study. One of the astronomers who fell under its spell was Peter van de Kamp.

Born in Kampen, The Netherlands, van de Kamp spent most of his life in America. A passionate classical musician who seriously contemplated a career in the performing arts, he instead indulged his other lifelong passion, that of astronomy. In spring 1937, he became director of Swarthmore College's Sproul Observatory in Pennsylvania. There he ran a programme to measure the positions of stars, looking for deviations known as parallax that could be used to calculate the star's distance from Earth. The nearer the star, the greater the angle of parallax that can be measured and so Barnard's Star was a natural target. In 1963, with a quarter of a century of observations under his belt, van der Kamp made the claim that Barnard's Star wobbled as it moved across the sky. Van de Kamp thought this movement resembled a corkscrew motion, which is exactly what would be expected if the star were being pulled by an unseen planet.

Working with photographic plates dating back to 1916, van de Kamp painstakingly measured the position of Barnard's Star over the years, and began to calculate what type of planet would be needed to make the star behave in this way. He went public at the April 1963 meeting of the American Astronomical Society, which took place in Tucson, Arizona. His first estimate was a giant planet 1.7 times the mass of Jupiter in an elliptical orbit with an orbital period of 25 years.

On 19 April, the *New York Times* reported the story with the headline 'Another Solar System is Found 36 Trillion Miles from the Sun'.

A few months later, van der Kamp revised this idea to state that there were two planets not one, each in a circular orbit. A planet 1.1 times the mass of Jupiter orbited every 26 years, while a planet 0.8 times the mass of Jupiter took its turn around Barnard's Star every 12 years. As he continued to take more measurements and hunt down older photographs, he continued to refine his estimates. In 1975 he announced revised figures of a 0.4 Jupiter-mass planet in a 22-year orbit, and a 1 Jupiter-mass planet in an orbit of 11.5 years.

Intrigued, others began looking for the planet too, and that's when the problems began.

Photographs from other telescopes did not show any wobble in Barnard's Star at all. Then another astronomer at Sproul Observatory looked again at the plates used by van der Kamp. John L. Hershey did indeed find the evidence for a wobble but not just in Barnard's Star. It was present in the dozen other stars he studied as well.

Either they all had planets, or something else was clouding the measurement. Hershey found that the largest wobbles corresponded to times when the telescope's lens had been removed and replaced during refurbishment. He concluded that van der Kamp was being fooled because he had failed to take these systematic errors into consideration, but the Dutchman refused to submit. Although the weight of opinion was fully against him, he continued to maintain that there were planets around Barnard's Star until his death on 18 May 1995. His stubbornness also helped fuel the prejudice against astronomers claiming to have detected planets. Just a few years before his death the astronomical community had received a new reminder of the dangers of making such claims.

World-renowned British astronomer Andrew Lyne worked at the Jodrell Bank Observatory in Manchester. He was an expert in analysing the radio signals from dead stellar cinders known as pulsars. These are objects that contain several times the mass of the Sun but squeezed into a volume no larger than about 20 kilometres across. They are the densest objects in the known universe and are the remains of a star's nuclear heart.

At the end of a massive star's life, it explodes in what astronomers call a supernova. In this cataclysmic event, most of the star is blasted into space where its gas can be incorporated into the next generation of stars and planets, but the spent nuclear furnace at its heart collapses from about the size of the Earth to the size of a moderate asteroid to create the pulsar.

Technically, it creates an object known as a neutron star, so called because the protons and electrons normally found inside atoms are crushed together to become tightly packed neutrons. The neutron star is set spinning at a furious pace, often dozens of times a second. This would be undetectable if it weren't for the fact that neutron stars emit beams of radio emission that sweep through space like lighthouse signals. If these happen to cross Earth, radio telescopes see them as pulsating flashes, hence the term pulsar. Most astronomers thought that, in the extraordinary violence of the supernova, any planets orbiting the star would be vaporized, but tantalizing hints that this might not be the case began to surface.

The clues came in the timing of the pulses. If a pulsar was fixed in space, rotating at an unvarying speed, then the pulses would arrive with complete punctuality. But a number of pulsars seemed to display some variation around the average. Could these be caused by planets? As with the putative planets around more normal stars, subsequent investigation destroyed any such claim. Principally this was because the more pulsars astronomers found, the more they became acquainted with their behaviour. This showed that some were not the steady clocks originally assumed. Something akin to earthquakes must be taking place inside the neutron stars, altering the distribution of their mass, which in turn affected the timing. But in 1991, one pulsar discovered by Lyne and his students began to stick out from the crowd.

It was known only by its catalogue name of PSR B1829-10. One of Lyne's graduate students became convinced that its timing was varying periodically. Lyne was reluctant at first to consider the possibility, thinking that it must be the normal timing noise issue but the pattern was all wrong for that. The idea of a periodic, repeating signal lodged in his mind because the variation was so slight that it could only be caused by a planet. If so, it would be a major discovery. He showed the student, named Setnam Shemar, the technique for extracting an unseen companion's mass and orbit from a set of timing variations, or residuals as they are often called.

Shemar came back with a beautiful fit to the residuals. The pulsar's timings could be completely understood if it were being pulled by a planet the size of Uranus, in an orbit the size of Venus. For a couple of months Lyne and Shemar reanalysed their data, looking for anything spurious that might be tripping them up. When they found nothing they wrote a paper claiming the detection of a planet near pulsar PSR B1829-10 and submitted it to *Nature*. It was published on 25 July 1991 with the front-cover headline 'First Planet

Outside Our Solar System'. This was no newspaper hype; this was the cover of the world's most prestigious science journal.

It was a bombshell because it would require a total rethink of supernova explosions and the hardiness of planets. Could this strange new world really have survived the cataclysm? Was it originally much larger and had it been stripped down to a bare kernel? Had it formed in the aftermath from some of the stellar debris?

Half the world away, in Puerto Rico in the Caribbean, Polish émigré Alex Wolszczan was as excited as everybody else to hear of this amazing object. He was also a little disappointed because he was working on some puzzling data of his own that implied that another pulsar, PSR B1257+12, had planets in orbit. The disappointment came because he had now been beaten to the first discovery of a planet beyond our solar system.

If anything, however, his planetary system was more interesting. Using his own data and other data supplied by Dale Frail, an American astronomer from the National Radio Astronomy Observatory in Socorro, New Mexico, he spent the whole summer and autumn trying to calculate a solution. Not one but two planets were needed to fit the data: one orbiting every 67 days, the other every 98, both in somewhat elliptical orbits. But the real news was that the planets were not large like Lyne's; both were just a few times the mass of the Earth. By December, Wolszczan began presenting his results at conferences. He and Lyne were invited to give back-to-back talks at the American Astronomical Society's meeting in Atlanta in the second week of January. That's when disaster befell Lyne.

He had made an early start to the new year and was in his office on 2 January 1992, poring through the data to refine the orbit. He noticed that the position of the pulsar was slightly different on some of the files. This was an oversight and he immediately corrected the figures to the more accurate reading, then he ran the computer software again to update the orbit of the planet. It took the computer about three minutes to do the calculation. What Lyne saw amazed him. The six-month orbital periodicity on the data that he and his students had been interpreting as a planet had completely vanished. It was a mistake brought about by using erroneous positions for the pulsar.

Suddenly, Lyne realized what he had to do. He flew to Atlanta and, at the appointed time, just before lunch, took the stage as 1,000 astronomers crushed into the hotel ballroom to hear about these fascinating worlds. Lyne talked the audience through the whole procedure. When he admitted the mistake, there was a collective gasp from the audience. Only a very few trusted souls had been clued-in to the bombshell the Englishman intended to drop. The planet was a phantom, introduced by a faulty analysis that assumed Earth's orbit was a perfect circle instead of a slight ellipse. It was an elementary mistake. 'Our embarrassment is unbounded,' Lyne concluded, visibly upset, 'and we are very sorry.'

The audience rose to its feet and gave him a standing ovation for the courage he had shown in admitting his mistake, and apologizing in person. In many ways, it was science at its best.

Wolszczan spoke next, telling the shell-shocked crowd that his planets continued to stand up to scrutiny. He speculated that they were second-generation planets, formed from the debris of the exploding star. This was because their pulsar was a special kind that spun extremely quickly, completing a full revolution in just 1.6 milliseconds. There was no way that astronomers could contrive their computer models to give birth to a pulsar spinning this furiously but it was possible to achieve the speed if matter from the destroyed star rained back down on the pulsar, spinning it up like a top. The laws of physics dictate that this matter would almost certainly first form a disc around the pulsar's equator. It was at least conceivable that the planets could have condensed from matter in this disc as the rest was falling onto the pulsar's surface.

Wolszczan and Frail's detection has stood up to the closest of scrutinies, and to this day remains the first detection of planets beyond the solar system. Originally referred to as extrasolar planets, from the Latin *extra* meaning outside, this term is now usually shortened to exoplanet.

At the time, it showed that the bona-fide detection of planets was not such a crazy idea after all. Nevertheless, Mayor and Queloz were feeling under pressure when they prepared to make their announcement about a planet circling 51 Pegasi.

Despite the paper having passed all the tests set for it, *Nature* seemed reluctant to publish. Weeks went by as the astronomers waited for news of when their work would be see the light of day. Nothing came from the editors. As October approached, Mayor decided to take matters back into his own hands. He and Queloz were attending a conference in Florence, Italy, dedicated to the discussion of 'cool stars, stellar subsystems and the sun'. It provided the perfect academic platform. Mayor contacted *Nature* to make sure he would not jeopardize the paper by announcing their result before publication, and he was told to go ahead on one condition: that he not mention that their paper was coming out in *Nature*. Clearly, the journal still had doubts.

There was no attempt to gather the press to Florence for the announcement, and when Mayor made his announcement there was a sense of disbelief among the astronomers in the room. He was postulating a planet with at least half the mass of Jupiter in an orbit of just 4.23 days. This meant it was travelling 136 kilometres every second under the influence of the star's gravity. In reverse, the planet moved the star at 70 metres per second, which was the signal that Queloz had captured in the spectroscope. To hold an orbit of that period, the planet itself would have to be just 7 million kilometres from its parent star. In our own solar system, Earth is 150 million kilometres from the Sun; even Mercury is 58 million kilometres from the incandescent surface.

Calculations showed that the planet would be fearsomely heated, and astronomers began referring to it as a 'hot Jupiter'. The gravity of the star would prevent the planet from spinning, so it would be forced to hold just one face to the star, just as the Earth's gravity holds the Moon so tightly that we see only one face of the Moon. The planet, dubbed 51 Pegasi b, was assumed to be gaseous like Jupiter, and the heat from the star would raise its temperature to a blistering 1,200 degrees Celsius. With one side of the planet fearsomely heated while the other always faced the cold expanse of space, there would be perpetual hurricane-force winds in its atmosphere, as the heat tried to equalize the planet's overall temperature.

It was no wonder that *New Scientist* reported the claim with a quote from Franco Pacini, director of the Arcetri Astrophysical Observatory in Florence, who attended Mayor's talk. He said, 'It is almost crazy.'[3]

Across the Atlantic, Geoff Marcy was equally disbelieving. He had been given the heads-up in an email from the highly respected physicist Philip Morrison, and thought to himself, 'Here we go again.' It was going to be another false claim to add to the string of previous embarrassments. Unlike most, who could only scratch their heads and wonder at the validity of the result, Marcy could do something about checking it. He had four nights of observing time at Lick Observatory scheduled only a week later, so he and Paul Butler junked their schedule and observed 51 Pegasi. They quickly saw that the star was indeed wobbling, and over those four nights saw the star complete almost a full pirouette as described by the Swiss team. It was the unmistakable signature of a planet. They drove away with mixed feelings. On the positive side, they were convinced that they had confirmed the first planet outside the solar system. On the negative, they had no idea how far ahead Mayor and Queloz were with their survey. In fact, the Europeans had focused so much of their effort on 51 Peg that they had neglected to look at other stars. Marcy and Butler had a clear field but didn't know it.

When they announced their confirmation of 51 Peg b, it sparked a storm of media interest. Within a week *Nature* had published the paper it had been sitting on, stoking more press attention. Queloz remembers that, for the next six months, he received so many requests from television, radio and the printed press for interviews and appearances that it was almost impossible to get any work done at all. In the States, Marcy and Butler were similarly in demand. This was particularly frustrating for Butler because he immediately wanted to analyse the data he and Marcy had collected over the last decade. They had believed the common wisdom and assumed that any planets close to stars would be right on the edge of detectability. Yet now they knew that if there were other tightly orbiting planets similar to 51 Pegasi b, each one would stand out head and shoulders above the noise.

After they found time to get to work, they found a planet around 47 Ursa Majoris, a Sun-like star some 46 light years away, within weeks. It was at least 2.5 times more massive than Jupiter, and took 2.95 years to orbit its star. If it could be magically transported into the solar system, it would sit part way between Mars and Jupiter, in the region inhabited by the asteroid belt.

A month later, another Sun-like star, 70 Virginis, yielded a planet. The star was being orbited by a heavyweight world of at least 7.5 Jupiter masses. The most important thing, however, was its orbit. It was highly elliptical whereas the other planets had been in circular orbits. To Marcy this was significant. A defining moment for astronomy had taken place in the early 17th century when the German mathematician Johannes Kepler had succeeded in describing the orbits of the planets using mathematics. It took him nearly a decade to analyse the movement of Mars across the night sky. At the end of this gargantuan effort, he described the movement of Mars and the other planets of the solar system in just three mathematical laws.

The first law is that planets move on elliptical orbits around the Sun. In this context, a circle is just a special case of an ellipse. The second is a mathematical rule for how a planet speeds up when it is close to the Sun and how it slows down when it is far away. The third law relates the average speed of a planet to the size of its orbit, showing that more distant planets move at a slower speed. It was these laws that Mayor, Queloz, Marcy and Butler were applying when they did their analysis. Whereas it was possible to imagine things happening on the star that might mimic a circular orbit – a pulsation say – it is much harder to conceive of something that could mimic the exact Doppler signal of an elliptical orbit.

As much as Marcy believed the data on the individual stars, there was always a shadow of a doubt about misinterpreting something. However, this trio of results, taken together, convinced him completely: 51 Peg b showed that there were planets out there approximately the size of Jupiter; the larger orbit of 47 Ursa Majoris b showed the link between hot Jupiters such as 51 Peg b and our solar system; and the highly elliptical orbit of 70 Virginis showed that these were real orbits.

However, not everyone was ready to agree.

THREE

The planet bonanza

Things became turbulent for the researchers, who were facing polarized responses. While the media were celebrating their discoveries and lauding them for their ingenuity, the astronomical community were sceptical. Some were even on the attack. This was no surprise. Every major discovery must withstand assault. It is the way that science is tested, a kind of intellectual trial by ordeal. Mayor and Queloz were receiving requests from journals to referee papers that pointedly sought to discredit their finding. Making time to answer these criticisms while simultaneously talking about their discovery to the press, let along searching for more planets, was a source of pressure.

In most cases, once they found the time to look at the criticisms, it was straightforward to present counter-arguments or more data to show that the concerns were groundless. They didn't take offence at these challenges. It was understandable that astronomers were sceptical because the discovery flew in the face of the accepted wisdom at the time. Yet one paper did leave a bitter taste.

It was to be published in *Nature*, the very same journal that had carried the original groundbreaking paper. Although the editors of the journal had the courtesy to send Mayor and Queloz the paper in advance, and even asked for their comments, they made it very clear that the Swiss astronomers were not being consulted as referees. In other words, no matter what data or explanations they provided, they would have no direct influence on *Nature*'s decision about whether or not to publish. As they read the paper, Mayor and Queloz realized that this was a stiff challenge indeed.

The paper had been written by a single author, David Gray, an astronomer from the University of Western Ontario, Canada. Like Marcy and Butler, he had built his career on studying the spectral lines of Sun-like stars, even writing a book on the subject that had become something of a standard text for astrophysics students across North America. Whereas Marcy and Butler had been concerned with the movement of spectral lines, Gray was interested in the way they can change their shape.

He had been studying 51 Pegasi b since 1989 and had concluded that the star was mildly variable. This was important for what was to come because if a star's brightness varies it must be pulsating. Those pulsations will involve movement of the star's surface in some way and that will impart a Doppler shift that will alter the shape of the spectral line.

Gray did not think too much about this at the time of 51 Pegasi b's discovery but about a year later he started to feel troubled by the bonanza of planets that were being announced and unquestioningly believed. In an interview for *Sky & Telescope* magazine in 1996, he described the sense of interest surrounding exoplanets as 'almost hysterical'.[4] So he set about conducting a reality check using his own data to see if the planet hunters were being fooled.

He gave his paper the provocative title, 'Absence of a

planetary signature in the spectra of the star 51 Pegasi'.[5] His argument rested on an analysis of his own data in which he saw that a spectral line associated with iron was changing its shape rather than moving wholesale. It appeared that the line was morphing with the period very close to the orbital period of the planet. So he concluded the summary with 'The presence of a planet is not required to explain the data.'

Mayor and Queloz immediately requested his data so that they could analyse it themselves to see if they could reproduce his signal. But their request produced no response. In the absence of the data, and in light of the strength of the criticism in the paper, they teamed up with Marcy and Butler, and then took an extraordinary step. They published a rebuttal directly on the Internet so that it appeared simultaneously with the publication of Gray's paper.

Unable to analyse Gray's data, all they could do was point out three reasons why they stood by their planet interpretation.

Firstly, an already published independent study of 51 Peg by astronomers in Texas had failed to replicate Gray's changing line shapes.

Secondly, the brightness of 51 Peg was steady to 2 parts in 10,000 according to work by an astronomer in Tennessee State University. Such stability was very peculiar if 51 Peg really was a variable star. All other variable stars changed their brightness.

Thirdly, the motion of the spectral line that they had deduced was a perfectly simple sine curve, whereas variations in a star would usually be expected to have a number of complications, like the harmonics produced in a musical instrument. These were completely absent in the data.

Initially Gray was unfazed, saying that he had been working on models of variable stars that could produce pulsations across the surface, which did not change the star's brightness. He also thought that since 51 Peg was older than the Sun, it might be behaving in a way that had yet to be anticipated by astronomers.

Both parties agreed that the route forward was more data, and thankfully new technology was coming online. Marcy and Butler were about to start observations at the huge Keck 10-metre telescope on Hawaii, and the independent Texas team who had failed to replicate Gray's findings were about to shift their spectroscope to an experimental 11-metre telescope called the Hobby–Eberly Telescope in western Texas. Gray too planned more observations and, by late 1997, he had enough new observations to draw a conclusion. He again submitted his work to *Nature*, who again decided to publish it.

It was printed on 8 January 1998. Gray had again chosen a blunt headline: 'A planetary companion for 51 Pegasi implied by absence of pulsation in the stellar spectra'. As only the most confident scientists can bring themselves to do, he had done a complete about-face in light of new data. This paper was essentially a retraction of his previous one. In it he said the planet was the best explanation because his own subsequent data had failed to show any change in the line profiles, only movement.

He finally supplied data to Queloz, who began to scrutinize the numbers. Soon Queloz found what he thought was the source of Gray's scepticism. It was a slight error in the original period of the planet. Instead of using the corrected one that had become apparent with more data, Gray had been analysing his data with the original quoted period. This meant he wasn't always looking at the centre of the spectral lines and so it looked as if the shape was changing.

Nevertheless, there were still many who would prefer that these exoplanets simply did not exist.

51 Pegasi b and the other exoplanets were a bitter pill for some astronomers to swallow. If you took the new discoveries at face value, they proved once again that nature is far more creative than human beings at imagining reality. The new planets – hot Jupiters, as they were called – shattered the neat and tidy hypothesis of how planets formed, based on the template of our own solar system.

There is no getting away from the fact that our collection of planets is an extremely orderly place. We have a sequence of four rocky worlds, Mercury, Venus, Earth and Mars, followed by the same number of gaseous planets: Jupiter, Saturn, Uranus and Neptune. The major difference between these two classes is their size: the rocky planets are relatively small, with Mercury just 4,879 kilometres across and Earth 12,742 kilometres. They are mostly made of rock and metal, so have a high average density, and are topped by a relatively thin atmosphere.

The bigger planets on the other hand are often referred to as gas giants. They are relatively large. Jupiter is the greatest at 139,822 kilometres in diameter, ten times the size of the Earth. The smallest of the gas giants, Uranus, is still four times wider than Earth at 50,724 kilometres.

The exquisite order of it all – smaller rocky planets close in, followed by larger gas giants further out – seemed to beg for an explanation, and in the 1980s astronomers found something to tie it all together. The Infrared Astronomical Satellite (IRAS) was launched at the beginning of 1983, the same year that Marcy joined the staff at San Francisco State University. Later that same year, IRAS discovered excessive amounts of infrared radiation coming from four nearby stars. By excessive, astronomers meant more than the star alone could be generating.

One of these infrared excess stars, the young star Beta Pictoris, became the subject of an investigation the next year by Bradford Smith, University of Arizona, and Richard Terrile, NASA's Jet Propulsion Laboratory. They observed Beta Pictoris from Chile's desert-bound Las Campanas Observatory, using a specially constructed mask to blot out the light from the central star and render any fainter surroundings more easily visible.

Nothing much showed up at the time but, back at their computers in California, a routine analysis showed there was something around the star. Appearing as two thin extensions of light on opposite sides of the star, it was exactly what a disc of dust and gas would look like if it just happened to be edge-on to our view. Importantly, cool dust could give off the large quantities of infrared radiation that would explain the IRAS observations. From the size of the disc on their image, Terrile and Smith estimated that it contained enough matter to build about 200 planets the size of Earth.

Given that the calculated age of Beta Pictoris is just 8–20 million years (whereas the Sun is 4.6 billion years old) this observation was excellent evidence that planets formed by condensing out of a disc of dust and gas. It also naturally suggested why our solar system was so orderly. To see why, we have to wind back more than 4.6 billion years to when the Sun and the planets of the solar system were forming in a disc of their own.

Back then, the young Sun was well developed and was already shining energy into space, heating the surrounding disc of matter. In the disc's inner regions, where the rocky planets formed, the temperatures were high enough to prevent anything but rocks and metal from condensing into solid dust. This explains the predominant composition of Earth and the other rocky planets. It also explains their size because planets grow from the matter that sits in their orbits. In the case of the inner planets, the orbits are relatively small, and so the amount of matter is restricted.

Further out, the temperature of the disc was lower and this allowed icy material to condense as well as rocks and metal, providing more material from which to build larger planets. Also, the larger orbits of these planets boosted the amount of matter available for construction and that led to big planets, so large that they could then generate the gravity needed to capture the most fleeting chemicals of all: gases. So Jupiter and the other gas giants were born.

Computer models could reproduce this behaviour beautifully. It all made so much sense: simple, elegant, yet – as 51 Pegasi b showed – wrong. There was just no way in the accepted scenario for a gas giant planet to form that close to the Sun. It gave some astronomers just the excuse they were looking for to revise their models of planet formation. The more they looked at the details, even for the solar system, the more things they were finding that just didn't add up.

As envisioned by astronomers, planet formation is not a gentle process. It involves the condensation of gases into solid particles of 'dust'. This is not the common-or-garden detritus that plagues our homes but usually solid bits of silicon and carbon. The other elements condense too to form minerals. This process leads to pebble-sized, boulder-sized and eventually mountain-sized objects called planetesimals. The asteroids in our solar system, which are mostly found between Mars and Jupiter, are essentially leftover planetesimals. When the planets were forming, the entire solar system was made of these planetary building blocks. They were continually colliding with each other, melting together and solidifying into Earth and its sister worlds.

Simulating the birth of the planets in this way shows that most could have formed in about 100 million years. But for Uranus and Neptune, there are some problems. They are situated 15 and 30 times further from the Sun than Earth. Out in these distant realms, the planet-forming disc was thinning out. Computer simulations show that this would have made it difficult for the two outer planets to reach their present size. To solve this conundrum, some researchers suggested that Uranus and Neptune had formed closer to the Sun and then been pushed outwards by the combined gravity of Jupiter and Saturn. Thus was born the concept of planetary migration.

According to this scenario, which is called the Nice model because it was originally developed at the Observatoire de la Côte d'Azur, the real engine of the migration was Jupiter. During the later stages of its formation, when most of its bulk was accumulated, Jupiter would continue to interact with the remaining planetesimals. Some of the encounters would be collisions, adding to the bulk of the planet; others would be near-misses, and these would exchange some gravitational energy with the planet. In the Nice model, the net result of these gravitational interactions was to set Jupiter on a spiral trajectory outwards. It was painstakingly slow but over a hundred million years or so, the giant planet would gradually move outwards. As it moved, even from hundreds of millions or billions of kilometres away, its gravity would push Saturn, Uranus and Neptune outwards too.

It is not only the formation of Uranus and Neptune that poses problems for theorists. Mars too is something of a puzzle. It formed beyond Earth, so had a larger orbit from which to feed and grow, yet it contains only a tenth of Earth's mass. What stunted its growth? According to a subsequent development of the Nice model, known as the Grand Track Hypothesis, the modifying influence was Jupiter again. In this scenario, somewhat different gravitational interactions with Jupiter set the giant world migrating inwards towards the Sun, exerting its gravitational influence in scattering planetesimals out of Mars' reach, arresting the growth of the red planet.

Jupiter may even have reached the current orbit of Mars before retreating again under the influence of Saturn's gravity. During these wanderings, the other planets moved too, first squeezing in and then spreading back out into the orbits we currently see them in. It is difficult to know how to discriminate between the Nice model and the Grand Track Hypothesis in our solar system but 51 Peg b certainly provided some form of corroboration that planetary migration could indeed take place – how else could you get a gas giant planet into such a tight orbit around a star?

Nevertheless, doubts reigned supreme. A gas giant almost skimming the surface of its star would have to have migrated a very long way from its place of formation. The theorists had no scenario by which this could happen; some even thought that it was flatly impossible and, in the face of such doubt, some found it simpler to dismiss the exoplanets altogether.

Queloz found it difficult to swallow the doubt that seemed to shroud his discovery. Even among those who did believe the work, there appeared a tendency to diminish the achievement as a flukey find – something that just popped out of the data unbidden. This has an eerie resonance with similar arguments that surrounded William Herschel when he discovered the planet Uranus in 1781. Astronomy at the time was dominated by the routine measurement of stellar positions to improve the navigation tables. Herschel broke from this tradition and conducted a survey of the night sky with a state-of-the-art telescope that he had constructed. His reward was to be the first person to recognize the seventh planet of the solar system. All the other planets known at the time had been recognized since antiquity because they are visible to the naked eye, so Herschel's discovery was epochal.

His detractors – mostly other astronomers – dismissed the find as mere luck. Herschel's supporters called those making such claims 'jealous barking puppies', and the man himself mounted a stout defence, claiming that he had 'gradually perused the great volume of the Author of Nature and was now come to the page which contained a seventh planet.'

Queloz developed a similar argument, saying that he discovered the planet because he did his job correctly. The key to this was making no assumptions about the nature of exoplanets. Instead, he let the data speak for itself. Despite the fact that theory had no explanation for the planet, the data was telling a different story. Fighting against his own reactions to the 'impossible' planet, Queloz developed a sense of what he calls 'pure science'. He describes it as where everything in you is screaming 'impossible' yet you fully believe your data. It was a transformative moment for him as a scientist. He learnt to cling to the data as though it was driftwood on a stormy ocean.

Despite the scepticism, by June 1996 the roster of exoplanets had grown to seven. Marcy and Butler had added another two. The first was a world containing at least 80 per cent of Jupiter's mass in a close orbit of 14.7 days around its parent star. The star itself is called 55 Cancri, and is similar to the Sun but forms part of a binary with another, smaller star. The second planet was a heavyweight at least 4 times more massive than Jupiter. It orbits the star tau Boötes, which is somewhat hotter and more massive than the Sun, and also part of a binary star system. A 'year' on this planet was just 3.3 days long.

The other two planets that had been reported used subtly different techniques. One planet was inferred through further studies of the disc around Beta Pictoris. Instead of being smoothly flat, the disc was warped, suggesting that a planet must be present.

The final suggestion proved to be the most controversial. American astronomer George Gatewood was working at the University of Pittsburgh. He claimed to have discovered evidence for two planets around a nearby star called Lalande 21185. He had not used the Doppler effect but instead had looked at the changing position of the star on a sequence of photographs. The careful measurement of a star's position is known as astrometry. It can reveal planets because the star moves in response to a planet's gravitational field, describing a small pirouette when a series of photographs are compared.

It was the astrometric method that Peter van der Kamp had used throughout the late 20th century to make his false claim of a planet around Barnard's Star. Back in the 1970s, Gatewood had taken his own images of Barnard's Star and used them to refute van der Kamp's claim. He had also taken images of Lalande 21185, and argued that there was no evidence for a planet around it either. Now the boot was on the other foot.

In 1996, Gatewood presented an analysis of 66 years' worth of images of the star and made a claim for not one but two planets in the system. He thought that the star was moving with such a complicated motion that only a pair of worlds could pull it that way. If true, it would be the first detection of a multi-planet system beyond our own. This was significant partly because the star was the fourth nearest to the Sun.

The press lapped it up, eagerly reporting the news. Gatewood calculated that one planet was larger than Jupiter and travelling in a 30-year orbit at a distance equivalent to Saturn's. The second planet was a bit smaller than Jupiter and could be found in a 5.8-year orbit, which would put it squarely in the asteroid belt between Mars and Jupiter in our own solar system.

Other astronomers were sceptical. It was notoriously difficult to measure astrometric positions to the accuracy that Gatewood was reporting. One way to perform a reality check was to see if there were any velocity changes in the star. The answer was no; Geoff Marcy had studied it already and found nothing. The star was so stable that he used it as a standard reference.

This in itself was not definitive because the velocity method is most sensitive to planets that are orbiting along our line of sight, whereas the astrometric method works best for planets orbiting face on. This is because a face-on planetary system allows the full pirouette of the central star to be seen by the camera. However, in this configuration there is no component of motion along the line of sight and so there is no Doppler effect. To acknowledge this, the exoplanet hunters refer to the velocity method as the radial-velocity method. This makes it clear to other astronomers that they can only detect movement in the radial direction, i.e. along the line of sight. Given this limitation, it was always possible that Lalande 21185 was entirely face-on, and so had no detectable radial velocity even though it was moving, but it was highly unlikely.

More doubt was cast when it was pointed out that, just four years earlier, Gatewood had released a paper categorically saying that there was no evidence for a planet in the system. Others looked and failed to see any evidence for planets, so most became sceptical. On the heels of David Gray's criticisms of Mayor and Queloz, any false claim was unhelpful to say the least. Marcy failed to secure some grant money for his research during this time, Mayor had a phone call from the Swiss science funding body asking him to explain again why he thought it was worth spending money on planetary detections. There was no doubt that Gray's paper had rattled many.

Marcy was invited to give a talk about these planets in Houston, Texas, the home of NASA's Mission Control and the University of Texas Lunar and Planetary Institute. When he got there, things were not as he expected. Instead of a lecture theatre, he was shown into a small room. Instead of an audience of students and academics, there were just half a dozen scientists. Instead of a talk, he was interrogated about his work. There was no trust in the room, just hard-nosed scepticism. It came as such a blow to Marcy's self-confidence that his old spectre of mental illness started to creep over the horizon. Much later, he told the *New York Times* that it sent him into a tailspin, that he 'was back to feeling stupid'.[6]

If only someone could take a picture of one of these planets, a spot of light to demonstrate that there was something real there after all. But it was a pipe dream; decades of technological development would have to pass before anyone could build a telescope capable of such a feat. It threatened to be an impasse.

Thankfully, Mother Nature had another trick up her sleeve.

Otto Struve's 1952 paper in which he suggested using the Doppler effect to find exoplanets had suggested another method too. He pointed out that some systems would be orientated in such a way that there would be an eclipse when the planet crossed the face of its central star. This would block out some of the star's light and the subsequent dimming could be measured to reveal the planet. These days astronomers call such events transits, preferring to reserve the word 'eclipse' for when the two aligned bodies appear to be almost identical sizes.

As well as being a detection method in their own right, transits also offer a way to corroborate velocity detections. While not all of the planetary systems will be aligned sufficiently to give rise to a transit, just as not every new moon creates a total eclipse, a few of the detected planets will cross in front of their star. It was just a case of continually looking until one was found.

In the few years after 1996, Marcy and Butler became an exoplanet discovery machine; they gained time on some of the largest telescopes in the world and built a team of planet hunters by attracting students and collaborators. In Europe, Mayor and Queloz pressed on with their planet search. Independent researchers also started to join the field. In 1999, as winter gripped the northern hemisphere, the exoplanet tally stood at 24. One of these was to become highly significant. It was HD 209458 b, discovered by Marcy and Butler.

The closer a planet hugs its central star the more likely we are to see it transit. Orbiting a Sun-like star in just 3.5 days, HD 209458 b was so close that the chances of a transit seemed high. It had a mass of at least 0.69 Jupiters, bringing it within range of extremely modest telescopes. With hindsight, the technology to make such detections had been available in the 1970s. If someone back then had thought to look, they could have made these discoveries. They didn't, but a number of observatories were now well equipped and on the lookout.

Working at the Keck Observatory, Hawaii, Marcy and his growing team discovered HD 209458 b on 5 November 1999. It was a Friday night, and the analysis software they had developed showed that there was the possibility of a transit coming up on Sunday. Marcy got on the phone to his collaborator Gregory Henry from Tennessee State University. As Marcy had done many times before by this stage, he told him of the possible transit and Henry agreed to search. At the appointed time, he fired up the remote telescopes he controlled in the Patagonian mountains of Arizona.

The transit would last for about three hours, and would begin just before the star set below the Arizona mountains. Sure enough, when Sunday evening rolled around, Henry saw the starlight drop by about 1 per cent. This was the ingress, the beginning of the transit. Although he did not see the egress, when the planet leaves the face of the Sun and the light levels return to normal, the ingress was enough to tell him that the shadow of a planet 150 light years away had touched the Earth. There could be no doubt any more. This was unimpeachable evidence that something the size of a planet was in orbit around the star. It was the next great discovery since 51 Peg b, and it sparked a race to claim the discovery.

The bone of contention was that Marcy had phoned not only Henry that night. Desperate to confirm whether these were planets or not, and fearful that clouds might obscure Henry's view, Marcy had also phoned astronomer David Charbonneau at Harvard University. Charbonneau and colleagues had been monitoring stars looking for transits but they had fallen behind on analysing their data. Marcy asked if they had looked at HD 209458 b. It turned out that they had observed the star a couple of months previously but had not analysed their data. When they did, the transits jumped out immediately. They had seen a complete crossing on 9 September and again on 16 September. With this data they calculated that the starlight had dropped by 1.7 per cent.

Now the race was on to publish the result. Within a few days Marcy's team had submitted for publication the velocity data, along with Henry's transit ingress data. Charbonneau had submitted his more complete transit data. Soon the world knew all about the detection and, after this announcement, Queloz noticed that things changed.

The doubts dropped away; some even felt the need to confess. They would wait until late evening, in bars at conferences and then turn to him with gravitas in their voice to say, 'You know, Didier, I never believed your work until the transits.'

It was the watershed. Now astronomers had two different strands to their planet-hunting bows. The velocity method continued as before, collecting data, publishing a steady stream of results. But the focus was shifting towards the transit method.

FOUR

Dreams of space

Even before the exoplanet detections, transits played an important part in the history of astronomy. Perhaps their biggest claim to fame is that they gave astronomers the first method by which to measure the exact distance to the Sun.

Kepler's 17th-century laws of planetary motion made it possible to relate the distances between the planets in the solar system to Earth's orbit. The distance between the Earth and the Sun was called the astronomical unit (AU). Although its absolute distance in miles was not known, Kepler's third law allowed distances in astronomical units to be calculated for the other planets, based on the time it took for Planet Earth to complete an orbit. Venus, with its orbit period of 0.615 years, has an orbital radius of 0.72 AU; Earth, with a period of 1 year, lies at 1 AU; Mars, with a period of 1.88 years, lies at 1.524 AU.

Being in a smaller and faster orbit, Venus offers the opportunity of a transit, which would allow astronomers to see its silhouette again the Sun. It passes between the Earth and the Sun every 583 days but a subtle misalignment of Earth's orbit with that of Venus means that a transit is not visible every time. The correct alignment only occurs every 122 years, followed by a repeat performance 8 years later, then another wait of 122 years.

The first recorded observation of a transit of Venus took place on 4 December 1639, when English astronomer Jeremiah Horrocks rigged a telescope to project an image of the Sun onto a board, and watched the black circle of Venus slip across the bright surface. Although it was little more than a scientific curiosity at the time, a change in thinking came in 1677, when the young Edmond Halley saw the other inner planet, Mercury, transit the Sun.

Halley was a young, ambitious astronomer who had quit his degree at Oxford in protest over what he considered was a backward syllabus. Winning royal patronage, he gained passage to the Atlantic island of Saint Helena where he began to measure a new star chart for the English Navy to use in navigation. Such an endeavour was of prime military significance at the time, yet Halley also found time for some pure science.

On 7 November 1677, he measured the silhouette of Mercury passing across the face of the Sun, and started to apply some geometry. It had been realized in 1663, by the Scottish mathematician James Gregory, that the time at which a transit began was dependent on the observer's place on Earth. Widely separated observers would be looking along different lines of sight and that would affect when the planet lined up with the Sun. If several could be timed accurately enough then some trigonometry would reveal the absolute distance to the Sun. It would be a major advance of astronomical knowledge to know the value of the astronomical unit in miles – to actually know how far away the Sun was. Although the observations would be difficult for the telescopes at the time, the mathematics was essentially an exercise in triangulation.

Astronomers identify four key moments in a transit, naming them contacts. First contact is the moment that the planet just 'touches' the bright edge of the Sun. Second contact is the moment that the planet is fully in front of the Sun's disc. Third contact is after the planet has crossed the face and begins to exit, and fourth contact is the moment that the planet's silhouette is finally lost from view because it has slipped fully off the disc.

Halley observed the transit of Mercury to test how accurately he could make such timings. Firstly, he would have to estimate precisely when the planet first cut into the Sun. This was difficult because the planet was too dim to be seen during its approach, so only when it covered part of the surface could it be divined. Then he timed when it was completely within the Sun and finally made two similar measurements as it touched the other limb, and disappeared completely.

In his makeshift observatory on the side of the Saint Helena hills, Halley met with limited success. He observed the transit from start to finish but Mercury was such a small dot that his measurements were too imprecise to furnish a good estimate of the Sun's distance. However it started him thinking. For decades, the idea lay in the back of his mind. Only in 1716 did he finally put pen to paper. Now facing old age, yet still four years before being crowned the second Astronomer Royal, he pointed out that Venus was a much better target for this work. It was nearer to Earth, and naturally larger, so its dark disc could be measured much more accurately. Halley wrote down the method by which transit timing could be turned into a distance using geometry. He then pointed out that astronomers would have to be patient because nature would next present this opportunity in 1761. Despite his advanced warning, it was not until 1760 that preparations for this extraordinary chance truly got underway. By then there was a sense of urgency and danger driving the proceedings.

Britain and its allies were at war with France and its supporters, fighting for control of foreign lands. It was a global conflict, perhaps the first 'world war'. Nevertheless, dozens of astronomers from across Europe were setting sail to make this measurement in the first large-scale international scientific collaboration the world had ever witnessed.

The chosen pair of British astronomers, Charles Mason and Jeremiah Dixon, knew that time was against them as their ship, HMS *Seahorse*, slipped out of Portsmouth harbour on 8 January 1761. They had six months to reach Bencoolen, known today as Bengkulu, a province of Sumatra, and they were in a Royal Navy vessel because the East India Company had said the voyage was otherwise impossible in the time available. The military solution also seemed a good choice given the ongoing hostilities. Should they arrive to discover the port in French hands, their orders were to divert to Batavia and observe from there. As the harbour dwindled behind them, little did they know that an encounter with the French lay not half a world away, but just two days in the future.

Seahorse was only 34 leagues out when she was engaged by *L'Aigrette*, a French frigate with nearly twice the firepower. The pitched battle lasted for an hour, culminating in a mutual withdraw at the sighting of a second British ship. Eleven of *Seahorse*'s complement were dead, 37 more were wounded and there was minor damage to the astronomical equipment. The ship returned home. The astronomers stepped back onto dry land and told their masters that was where they fully intended to stay. But they were no longer masters of their own destiny.

After repairs to the ship and a stern reminder from the Royal Society in London about the importance of bravery, the terrified astronomers were sent out again. There was simply too much science at stake for lily livers. If they did not make their observations, the chance to measure this critical event would be lost for the next eight years and now time was fully against them.

The further they could get, the better the accuracy, but it was impossible to make Bencoolen in time, and no way to observe from the pitching deck of the ship. They got as far as Cape Town and began their preparations.

On 6 June 1761, as predicted, Venus cast its shadow towards the Earth, and observers everywhere timed how long it took to cross the fiery disc. It was anything but easy. An unexpected trouble occurred at second contact. The silhouette seemed to swim about: one minute it would look like it was still clinging to the edge of the Sun, the next it would appear to jump fully inside the fiery disc, robbing observers of the precise moment to note the time. This optical illusion was coined the 'black drop' effect, and it ruined the accuracy of the measurements. The observers set about calculating the solar distance and all came up with different figures. All too soon, the astronomers were arguing across the English Channel, mirroring the war taking place in its waters. The black drop had stymied them all. On the upside, they had eight years to prepare and try again.

In 1769, a second series of expeditions took to the seas. Now safely in peacetime, and better prepared, these observers were able to take much more accurate measurements of the transit. The results flooded in, allowing the French astronomer Jérôme Lalande to combine all the data and derive a distance to the Sun of 153 million kilometres. Today, the accepted value is 149.6 million kilometres.

In 1999, 230 years later, astronomers were again excited about using transits to extract key physical data, this time about exoplanets. Once again, it was all down to timing. The period between first and second contact, known as ingress, depends upon the size of the planet. Larger planets will obviously take longer to ingress than smaller ones. The same is true for egress, between third and fourth contact. Only when the planet is fully in front of the disc does the light level drop by its maximum amount. So, as the exoplanet crosses the edge of the star (also known as the limb), astronomers can time the drop and calculate the planet's diameter and volume. If the exoplanet is also detected using the radial-velocity method, then its mass will be known too, and this means that the average density of the planet can be calculated using the volume. This gives astronomers a strong indication of whether the planet is rocky or gaseous. In the case of the first transit detection planet, HD 209458 b, the average density was about a third the density of water. Clearly the planet had to be made mostly of gas.

There were other things that transits could be used for too, such as detecting exoplanets much smaller than the giant ones seen by radial-velocity methods. One person gripped by such thoughts was NASA space scientist William Borucki.

William Borucki grew up in the farmlands of Wisconsin. With little in the flat landscape to capture his attention, it was the big sky above that called to him. In the 1950s, as the space race was taking off, he and his friends built homemade rockets. The local sheriff would close the road so that they could launch. It was more for show than safety, Borucki recalls. There was so little around that 'the worst that could happen is that the rocket might hit a cow on the way down'.[7]

Borucki's passion later took him to NASA's Ames Research Center in the heart of Silicon Valley, California, where he became a space scientist studying the hostile environment that greeted a spacecraft as it re-entered Earth's atmosphere. He developed the spectroscope necessary to analyse this and soon he was working on the largest rocket of them all: the 111-metre-tall Saturn V, which took men to the Moon. His scientific studies helped drive the design of the heat shield that would protect the astronauts as they hit the Earth's atmosphere at 11.2 kilometres per second on their return. He remembers that, at the time, he felt no anxiety about having got it right. The numbers all worked, the testing had all been done. There was no doubt in his mind that the heat shield would work: the numbers said so.

Long before the exoplanet detection successes, there was another study in the 1980s whose numbers caught his eye. It was to do with the technological requirements of detecting transiting exoplanets in order to answer the question of how many other planets like Earth there are in the Galaxy. Astronomers call such worlds Earth analogues, though the media and general public often refer to them simply as Earth's twins.

Sitting down to calculate, Borucki realized that tens of thousands of stars would need to be monitored simultaneously in order to stand a reasonable chance of catching such a planetary needle in the galactic haystack. There were sufficiently dense star fields in the Milky Way where this could be done but although it was possible to detect Jupiter-sized transiting planets using ground-based telescopes, Borucki quickly realized that to catch an Earth-sized world would require a space telescope. This is because the turbulence in Earth's atmosphere, which makes the stars twinkle, would mask the minuscule drop in brightness caused by the silhouette of an Earth analogue planet.

Measuring the brightness of stars is a discipline known as photometry. Astronomers still use a system pioneered by an ancient Greek astronomer Hipparchus of Nicaea, who lived between 190 and 120 BC. Long before the age of telescopes and detectors, Hipparchus ranked the stars visible from the Mediterranean into six classes. He categorized the brightest as first magnitude and those just visible to the naked eye as sixth magnitude. This approximate system survived until 1856, when English astronomer Norman Pogson placed the magnitude scale on a more mathematical footing. He proposed that each magnitude class should be separated by a factor of 2.5. With this rule, it means that a star of the 2nd magnitude was 2.5 times brighter than one of the 3rd, which itself was 2.5 times brighter than a 4th magnitude star. In this way, there was a factor of 100 between the brightest stars and the faintest that could be seen with the naked eye.

To catch an Earth analogue transiting a Sun-like star would require detecting a drop in the starlight of just 0.01 per cent or about one thousandth of a magnitude. To reach consensus on whether this was even possible with current technology, Borucki persuaded the NASA Ames Research Center to fund a number of workshops to bring experts together to discuss the issue. When these discussions were successfully concluded, Borucki won further funding to build two experimental photometers, one of which worked in an encouraging way, and one of which did not. But the failure was of no consequence; they only needed one method to work, and one looked highly promising.

In 1992, NASA announced a new sequence of missions, called Discovery, that would investigate planets. Mostly these were envisaged to be the planets of our solar system but missions to search for exoplanets were not ruled out. To start the process, NASA asked for proposals that they could study further. Borucki pulled together a team and proposed FRESIP (FRequency of Earth-Sized Planets). After a lengthy review, the NASA panel ruled that FRESIP was not ready to proceed because it was still unclear to them that the necessary photometer could be constructed with enough accuracy to detect Earth-sized planets. The panel lauded the proposal for its science ambition, however, and this spurred the team on to continue their efforts.

In 1994, the first opportunity to compete for funding to actually build a Discovery mission was announced. Borucki updated the FRESIP proposal with his team's latest work and submitted. The photometers again were the stumbling point. The proposal was rejected on the grounds that the photometer would be too expensive to develop and build for the money NASA was willing to commit within the Discovery programme. But the agency did award more money for more lab tests.

With this funding, Borucki showed during 1995 that a photometer of exactly the required precision was possible using a form of digital imaging known as charge-coupled devices (CCDs). These electronic sensors are dozens of times better at registering light than even the best photographic film, so are also much more sensitive to faint changes. Others on the team rethought the orbit for the space telescope, placing it around

Earth instead of far away in space, and this significantly reduced the overall mission cost. So when a new Discovery mission opportunity arose in 1996, Borucki submitted again. This time there was no quibble about the cost, but he was turned down because no one had demonstrated that it was possible to monitor tens of thousands of stars simultaneously. Yet the door wasn't completely closed. Again, a small grant was given for Borucki to build something to demonstrate this ability.

He struck a deal with the Lick Observatory, California, where Marcy and Butler were diligently discovering more and more exoplanets, and built a photometer to be used on one of the other telescopes there. Sure enough, the telescope and the photometer began to track the brightness variation in 6,000 stars, all visible in a single field of view of the telescope. Surely, now, Borucki and his team had jumped through all the hoops that NASA required. Times and attitudes were also changing. By now, the success of Mayor and Queloz, and Marcy and Butler, had placed exoplanets on everyone's lips. This was hot science, capturing public and academic attention – and Borucki had a ready-made mission to capitalize on this moment of history and interest.

To shake off the stigma of the previous rejections, the proposal team renamed the mission. Instead of the coldly technical acronym FRESIP, they called the mission Kepler to honour the great 17th-century German astronomer, whose mathematical laws of planetary motion they would use to analyse their results (see Chapter Two). The ploy did not work. In 1998, the mission was rejected again.

It was as if the review team were looking for reasons to reject rather than accept. The latest issue was that there was no proof that a photometer of the precision needed to detect another Earth could be built and operated under the conditions that would be encountered in a spacecraft. For example, the spacecraft would inevitably jitter a little as it pointed to the target star field; could the photometer cope with the noise this would imprint on the data or would it wash out the signal?

A less stubborn man would have turned away, but Borucki believes in numbers. If he could build a simulation of the conditions on a spacecraft and measure the operation of the photometer under those conditions then surely, he thought, the numbers would speak for themselves.

The rig he and a fellow Wisconsin scientist named David Koch built in the lab had mechanical actuators to vibrate the equipment to simulate the pointing jitter, and a drilled metal plate to simulate the star field. Each hole represented a star; bigger holes transmitted more light and so simulated brighter stars. They drilled holes at random and then lay a flattened wire across the centre of some. When they passed an electrical current through the wire it expanded by a tiny amount, blocking just 0.01 per cent more light – exactly the amount they needed for an Earth-analogue world detection.

The simulations were a complete success. Also, by this time, the first transit of a large planet had been discovered thanks to David Charbonneau (see Chapter Three). There could be no doubt that Kepler was a viable mission and in December 2001, after 17 years of dogged technological development, the NASA hierarchy finally agreed. Kepler was chosen to become the fourth Discovery-class mission, and construction began in 2002.

In Poland around this time, Professor Andrzej Udalski of the University of Warsaw was looking critically at an observational project he was masterminding and wondering if it could be repurposed. The Optical Gravitational Lensing Experiment, shortened rather appropriately for an astronomical campaign to OGLE, was originally designed to look for dark matter, the extra matter that most astronomers think exist in the universe.

Of the two possibilities for this dark matter, the subatomic particles are referred to as WIMPs (weakly interacting massive particles), while the dim celestial objects are named MACHOs (massive compact halo objects), with 'halo' being the astronomical term for the space surrounding a galaxy.

OGLE was designed to look for MACHOs, which would show up because gravity bends light. Calculations make it clear that as a MACHO drifts across our line of sight to a distant star, its gravity would focus the starlight and cause the star to temporarily appear brighter by 1–2 per cent. To spot these 'microlensing' events, Udalski designed and built a highly sensitive camera for use at the Las Campanas Observatory in Chile. In the wake of the transit detection of HD 209458 b, he realized that the camera was also sensitive enough to register exoplanet transits – all that was needed was computer code to look for dimming rather than brightening.

By 2002, the OGLE astronomers had identified 59 stars[8] that might be dimming in response to the passage of an attendant planet with estimated diameters of between 1 and 4 times that of Jupiter. To be absolutely sure of these detections, what was needed was a radial-velocity measurement to check if the stars were wobbling. Because all the stars were far away, typically about 5,000 light years, a large telescope was needed to make the measurements. Another Polish astronomer, Maciej Konacki, led a small team of astronomers, including one from the University of California, who won time on the Keck telescope. Of the 59 target exoplanets, 20 were discounted after closer examination: 8 showed clearly that they were stars in orbit, not planets, 1 was a duplicate entry, 7 were too faint even for the Keck, and 4 had been triggered by only one transit so could not be used to make a prediction about the planet's orbital period.

On their first Keck run from late June to mid July of that year, more fell by the wayside. Of the 39 remaining candidates, 25 were clearly being circled by faint stars, not planets. Of those that remained on the list, more observations showed that OGLE-TR-56 b stood out because the radial-velocity signal was clear and matched[9] the orbital period derived from the transit data. Indeed, it was a hot Jupiter planet. Subsequent follow-up observations also revealed that OGLE-TR-10 b was another hot Jupiter. Although just two planets out of 59 candidates was a poor conversion rate, it was confirmation that the transit method could make detections of its own rather than just confirm radial-velocity finds, as had been the case with HD 209458 b.

The transit method also revealed that there were real differences in the physical conditions across the hot Jupiters. By calculating the average density, OGLE-TR-10 b, like HD 209458 b before it, was found to be an extremely low-density planet, much lower than the gas giant planets in our solar system. However, OGLE-TR-56 b had a density twice as large as its peers, pegging it close to that of Saturn. It was a tantalizing glimpse at the differences and similarities that were to be found among these new-found worlds.

There was undoubtedly great science to be had characterizing these planets as individuals. They could tell us how representative the planets of our own solar system were of planets as a whole. No wonder then that NASA was not the only space agency interested in a mission to find transiting exoplanets. The European Space Agency was on the trail too – even if its version did not start out as being about planets.

British astrophysicist Ian Roxburgh was interested in the way that stars quake. This curious phenomenon had been known since the 1960s when the ingenious Robert B. Leighton, an astronomer at the Californian Institute of Technology, had used the Doppler shift to study the motion of the Sun's surface. He discovered that patches on the surface move up and down taking about five minutes to complete an oscillation. When one patch was moving up, another could be moving down. There were hundreds or thousands of such patches on the Sun, all keeping their own time, but no one knew why.

An extensive theoretical effort provided the explanation. The oscillations are caused by sound waves trapped inside the Sun. The sound waves themselves are generated by the turbulent motion of the interior. When they arrive at the surface, the wave pushes that patch of the surface upwards before being reflected back down. As the wave travels deeper into the Sun, it is deflected by the ever-growing density back round and up to the surface again, where the cycle repeats.

Astronomers realized that the pattern of these oscillations holds information about the internal structure of the Sun. The rate at which the density increases determines how quickly the wave is turned back up to the surface. So, by studying these rippling surface patterns, astronomers could understand more about the Sun's make-up.

By the 1980s, efforts to do this with the Sun were well advanced, and Roxburgh and colleagues had developed the ambition to do this for other stars as well. The challenge was that the turbulence of Earth's atmosphere would overwhelm the signals from stars which are much more distant than the Sun. To be successful, the scientists would have to send their equipment into space.

At the time Roxburgh was serving on a panel for the European Space Agency (ESA), charged with defining the agency's science priorities for the turn of the millennium. The programme was called Horizon 2000. He managed to persuade the others on the panel that a mission to investigate asteroseismology was something the agency should pursue.

In 1985, ESA began considering a potential mission dubbed PRISMA, for Probing Rotation and the Interiors of Stars with Microvariability and Activity.[10] After a preliminary study, known as Phase A, PRISMA was rejected because the technology to perform asteroseismology was judged to be insufficiently mature. In 1988, Roxburgh's team tried again. A new phase A study was initiated and lasted three years. This time, the conclusion was that the mission was technically feasible. The final selection would therefore depend on the relative merits of the science it could perform but, in May 1993, ESA's science programme committee decided not to build PRISMA. They judged that the science from a gamma-ray observatory would be more scientifically important at that time, and so a spacecraft called Integral was manufactured and launched.

Almost immediately, ESA put out a new request for mission proposals, so the team resubmitted an expanded and improved proposal based on all the work that had been done during the PRISMA study. They renamed the mission STARS (Seismic Telescope for Astrophysical Research from Space) and attracted more signatories to the proposal. Then, a very important thing happened: ESA asked Swedish astronomer Malcolm Fridlund to perform a feasibility study. Fridlund had become a staff member at ESA a few years after completing his Ph.D. at Stockholm University in 1987. His key insight was to see that a space telescope capable of performing asteroseismology would also be capable of detecting the drop in starlight resulting from a transiting planet.

He began to invite astronomers who had studied these possibilities to advise the STARS science team but then fell seriously ill with blood poisoning. He was replaced as study scientist by Italian Fabio Favata, who enthusiastically took on the project of merging asteroseismology and planet detection into a single mission.

At the time, about seven years before Mayor and Queloz's detection of 51 Pegasi b, planet detection was considered a long shot because of the time it was thought would be needed to confirm a planet. It all came down to the assumption that other planetary systems would more or less follow the pattern of our own. If we were looking back at the Sun from a distant realm, we would see Jupiter cause a transit only once every 12 years because of the size of its orbit. More distant Saturn would slip across the Sun just once in 30 years. And when you see a transit, you would ideally need to wait for a second identical one to know the size of the orbit. Then you could predict the next transit to see if you were correct. At that point, you could be certain that you were seeing a planet. To do that with Jupiter would need a mission duration of around three decades; to detect Saturn with certainty by this method would take the best part of a century. These were clearly impossible timescales, so astronomers switched their focus to smaller rocky planets, which in our solar system orbit much more quickly.

Just a few years would be needed to detect planets the size of Venus or Earth. However, to see these smaller planets meant building a larger space telescope. So the team wrestled with a design that was big enough to stand a good chance of seeing Earth-sized planets, yet was small enough to be built on the budget ESA had available for a medium-class mission, around €350 million.

That changed the moment that 51 Pegasi b was discovered and it became clear that other solar systems did not all follow ours in their architecture. The hot Jupiter planets were large and would transit every few days if seen from the correct angle. The discovery came within six months of the proposal deadline and the STARS team worked hard to pull everything together into a single coherent package for an April 1996 submission. Again they were disappointed.

They lost to a mission that was subsequently named Planck, by one vote. Planck was designed to study the leftover radiation from the Big Bang, and in 2013 reported results that appear to challenge the standard theory of how the universe began.

No matter how important the Planck mission was destined to be, at the time Roxburgh found the rejection of STARS depressing. They had come so close. What more could they possibly do? This was just the first knock-back for Favata, however, and he persuaded the team to keep faith and to try again. All that was needed was a new call for proposals, and ESA would be bound to issue one sooner or later.

Meanwhile, a smaller proposal that members of the PRISMA and STARS team had been working on was bearing fruit. It was an instrument rather than a spacecraft. Called EVRIS, the team had negotiated for it to hitch a ride on a Russian mission that was going to study Mars. EVRIS was a small space-borne telescope that would look for oscillations on ten of the nearest stars during the nine-month cruise to the red planet. Named Mars '96, the does-what-it-says-on-the-tin mission was launched on 16 November 1996 and arrived perfectly in its parking orbit around the Earth, but the rocket burn to send it on its way to Mars never happened. A malfunction stranded the craft and as a result it fell back to Earth.

It was a dismaying turn of events for Roxburgh. The only light at the end of the tunnel was that, at ESA, Favata was as enthusiastic as ever and, in 2000, a new call for mission proposals was issued. Everything was ready to submit; all that was needed was a great name for the mission.

Roxburgh was on a plane coming back from a meeting with Favata and the science team held at ESA's European Space Research and Technology Centre (ESTEC) in Noordwijk, The Netherlands. It is a short flight back to London, just 45 minutes, but it was enough for Roxburgh to think through the issue. It seemed to him that the missions ESA was favouring were the cosmological ones, which seek to understand where the universe came from and which study the most distant celestial objects. While it was the exoplanets that were capturing the public's attention, and stellar astrophysics that interested Roxburgh the most, it seemed to him that the wider community of astronomers were begrudging of anything that didn't further the discipline of cosmology. That's when he came up with the name of Eddington.

In 1926, Eddington had written *The Internal Constitution of Stars* in which he had penned the hope that 'it is reasonable to hope that in the not too distant future we shall be competent to understand so simple a thing as a star'. He was one of the architects of modern astrophysics, yet his name was equally well known to cosmologists as he had led the 1919 expedition that verified Albert Einstein's General Theory of Relativity, the principal theoretical tool of cosmologists. So, Roxburgh thought to himself, perhaps calling the mission Eddington would make it sound more cosmological, and therefore more palatable to ESA.

He was partially successful. ESA's science programme committee chose to contribute instruments to NASA's Next Generation Space Telescope, the successor to the Hubble Space Telescope, rather than build Eddington, but they did not fully reject the mission. They requested that work continue on how to build the mission and what science could be done. They placed the mission on the reserve list, promising that if more money became available then the mission would fly.

The Next Generation Space Telescope was originally planned for a 2009 launch but, in America, the ambitious telescope itself was proving far more difficult to build than anyone anticipated. Significant delays began to creep into the programme and it soon became apparent that the launch would have to be pushed back by years. As well as instruments, ESA had also committed to launching the mission. The delays meant that the cost of supplying the rocket also slid further into the future, so the agency had more money available than it first thought. In 2002, as NASA were starting work on building Kepler, ESA decided to use its spare cash to begin implementing Eddington.

The race was on – and it wasn't just to get into space.

FIVE

The race to the bottom

Astronomer Steven Vogt from the University of California, Santa Cruz, remembers looking at the data when 51 Peg b and the other early planet discoveries were announced and thinking that, as momentous as they were, the name of the game was not about finding hot Jupiters but about finding Earth analogues. He thought of it as a race to the bottom, but it was no easy feat. The spectroscopes needed to be at least ten times more accurate than they were.

Geoff Marcy knew it too. Vogt had been his Ph.D. supervisor at the University of California, Santa Cruz. The pair had not only stayed in touch but were colleagues using the Lick Observatory. Vogt had built the spectrometer, known as the Hamilton, that Marcy had used for his thesis and was now using with Butler to detect exoplanets. Long before the discovery of 51 Peg b, Marcy had been 'complaining' to him that the Hamilton spectrometer was not precise enough for what he really wanted to do: find Earths. Vogt agreed but there was not much he could do. He knew exactly how he would modify the design to make the spectrometer more precise but there was neither the technology nor resources at Lick to do the job. And besides at the time he had been preparing to build a new spectrometer for the mighty Keck Observatory that was springing up on Hawaii.

The Keck Observatory is an impressive sight. Situated on the summit of dormant volcano, Mauna Kea, Hawaii, it consists of two telescopes, each with a 10-metre-diameter mirror. In the 1990s, such mirrors were impossible to build as single slabs of glass and so each giant eye consists of 36 hexagonal mirror segments, fitted together and held in place by computer-controlled supports. The gleaming white domes that house these telescopes are two imposing eight-storey-high buildings.

The telescopes were funded by the W.M. Keck Foundation, a philanthropic organization founded in 1954 by Californian oil businessman William Myron Keck. He wished the foundation to fund scientific and other projects that would provide far-reaching benefits for humanity. By the 1990s, a pair of 10-metre telescopes working side by side was near the top of the astronomy wish list.

The first telescope began science observations in 1993, with its twin coming online in 1996. The project was managed by the University of California, hence Vogt's involvement, and the Californian Institute of Technology. Vogt's job was to design and build a world-class spectrometer for use at this world-class telescope. It was dubbed HIRES (High Resolution Echelle Spectrometer) and although he designed it to be a completely general instrument for use in all branches of astronomy, he realized that it would make a powerful planet-hunting device.

This was good and bad news for Marcy. For the first time in the endeavour, his decision to work from the comfort of a smaller university was proving something of a hindrance. True, it had provided intellectual succour in the early years of the project but now he really wanted time on Keck with HIRES. The difficulty was that, to be awarded time, you had to be a professor at the University of California.

Vogt came to the rescue. He was a UC professor, and he made Marcy and Butler an offer. The three of them should join forces. If Marcy and Butler agreed to build an iodine cell that would allow their detection software to work, Vogt would incorporate the device into HIRES. Straight off the bat, they could measure stellar velocities to just a few metres per second. Although this was not quite enough to look for Earth analogues, the size of the mighty Kecks would allow them to look at fainter and more distant stars then ever before. As part of the team, Vogt would devote all of his Keck observing time to searching for planets. Although he was involved in many other astronomy projects, all of which could have benefited from observations with a 10-metre telescope, he recognized that fate had placed him at the perfect spot for finding planets. How could he not do everything in his power to further such a historic moment in human science? How could Marcy and Butler refuse his help?

The trio set to work immediately and began to dream of turning Keck into a major planet-finding machine. The problem was gaining enough telescope time to do this. Vogt could get around 5 or 10 nights a year but this was not nearly enough to mount the kind of planet-finding effort they all had in mind.

Then, in 1996, everything fell into place. NASA bought into Keck, paying for a one-sixth share of the telescopes. The space agency wanted to search for planets as a way to repair its tarnished reputation.

*

NASA in the 1990s was an embattled organization, under attack for its lack of direction. The heyday of the Moon landings lay more than 20 years in the past, yet it was still the achievement that defined the agency. In 1989, on the 20th anniversary of the 20 July Apollo 11 Moon landing, President George Bush had stood on the steps of the Smithsonian's National Air and Space Museum, with astronauts Neil Armstrong, Edwin 'Buzz' Aldrin and Michael Collins sitting behind him and called on NASA to reignite the exploration of space. He called for the construction of an American space station, the return of humans to the Moon, and a voyage to Mars. He said that it was humanity's destiny to explore, and America's destiny to lead.

It was to be the rebirth of America's space ambition, and NASA wasted no time in putting together its plans. By November that year, NASA had a road map to achieve the President's requests. All it would take was $500 billion. Half a trillion dollars! It was an eye-watering amount, even if it were to be spread over a number of decades. It was obviously an impossible ask, but without such a long-term goal what did NASA actually stand for?

In 1990, a presidential report recommended that the agency should concentrate its efforts on the observation of Earth and the surrounding universe. The poster child for this should have been the Hubble Space Telescope, launched in April of that year, but its very first observations revealed that the $1 billion project was suffering from a terrible mistake: the mirror had been made to the wrong shape. That was just one of the low points.

There was also the space shuttle, which was ferrying astronauts to and from Earth orbit, yet seemed to have little else to do. By 1992, not a single module of the promised American space station had been launched because of budget overruns. NASA was so deeply in crisis that there was the smell of death about the organization. Something had to change.

The change came in the form of Daniel Goldin. In 1962, he had begun his career at NASA as a new graduate, studying propulsion mechanisms for crewed spacecraft. Now, in the dying days of George Bush's presidency, he was appointed as NASA Administrator, the top job at the agency, and set about finding achievable, affordable and inspirational projects.

He presided over the first servicing mission to the Hubble Space Telescope, which repaired the optical flaws and brought the observatory up to its design specification. Under direction from the newly elected President Bill Clinton, he negotiated international partnerships with four other space agencies (Roscosmos in Russia, ESA in Europe, JAXA in Japan and CSA in Canada) to turn the American Freedom Space Station into the International Space Station that continues to fly today. Next, he wanted something that would sum up the agency in a single breathtaking image, in the same way that the classic 'Earthrise' photo of our world appearing over the Moon's horizon had done during the 1960s and 1970s.

Earthrise had been captured on Christmas Eve 1968, from the Apollo 8 command module. Inside were astronauts Frank Borman, Jim Lovell and William Anders. They were on their fourth orbit round the Moon, the first humans to make it this far into space, and Anders was taking images of the cratered surface with a handheld camera.

'Alright, we're gonna roll,' said Borman, announcing the beginning of a routine manoeuvre. The onboard tape recorders captured the puff of a thruster, like a sudden sigh, and the spacecraft began to gently rotate. That was when Anders called out, 'Oh my God, look at that picture over there.' Only a few moments earlier he'd been wondering whether an impact crater on the lunar surface was actually volcanic but the roll was bringing something altogether more breathtaking into view.

'There's the Earth coming up,' he said with excitement in his voice. On the silvery limb of the Moon, the brilliant blue planet Earth was just rising into the sky. Distance had rendered our mighty world a fragile orb. Although home to around 3.5 billion people at the time, any one of the astronauts could eclipse it by raising nothing more than their thumb.

'Wow, is that pretty,' said Anders, lifting his camera to capture the moment.

'Hey don't take that, it's not scheduled,' joked Borman.

Anders snapped a black and white photo and then called to Lovell to pass him a colour film cartridge. When it arrived a few moments later, he loaded it into the camera and took the iconic picture: Earthrise. Even now it remains a popular talisman of the exploration of space and can be bought on posters, mugs and T-shirts. As clearly as it symbolized NASA's early achievements, the fact that it had not been replaced by the 1990s seemed to also highlight the agency's stagnation.

So Goldin wanted a new image that could speak of NASA's current achievement and pack as much of an emotional punch as Earthrise. The discovery of exoplanets, and the public interest in them, gave him the perfect idea. He wanted an image of Earth's twin planet. No one knew where such a planet was located, or even if it existed, but with more and more exoplanets being discovered all the time, most astronomers were convinced it was only a matter of time before a twin turned up, and Goldin wanted to dedicate NASA to taking its portrait. But first they had to find it, and that was where Keck came in to blaze a trail that would lead to the discovery.

As soon as NASA bought into the telescope, Marcy, Butler and Vogt wasted no time in getting their proposal together. They had it all: a proven technique and all the hardware ready to go. NASA agreed and granted them time. Together with Vogt's own 5-10 nights they could observe for 20–30 nights a year. That was enough to run a concerted planet search. And it got better. To build the HIRES spectrometer for the Keck, Vogt's lab at Santa Cruz had been re-equipped with the very latest optical equipment.

The limiting factor on the Hamilton spectrometer at the Lick Observatory was a lens known as 'the corrector plate'. In the Hamilton, it was curved the same way in all directions and that was a problem, because the light paths through different parts of the lens needed different levels of correction. Vogt could use the new equipment from HIRES to remanufacture the Hamilton's optics to include a more complexly figured corrector plate.

With the new lens in place, the team could reach an accuracy of 3 metres per second, more than a three times improvement on their original accuracy, and almost the same as they were getting with Keck. They drew up a list of 500–1,000 target stars and began a decade of extraordinary achievement. More and more planets fell at their feet, and in 1999, Marcy finally moved from San Francisco State University to become professor of Astronomy at Berkeley. As this was a campus of the

University of California, he also now became eligible for Keck nights, and the programme expanded again.

With so many nights spread across the year, they became regular commuters to Hawaii. In the early years of the search, they would drive a 4×4 up the winding cinder track to the top of the dormant volcano, more than 4 kilometres into the sky. Up in the noticeably thin air, they would spend the long nights just metres from the giant telescopes, combing the nearby stars for planets. Later, it was possible to remotely control the behemoths from the sea-level town of Waimea, saving the astronomers from having to cope with the rigours of altitude. Vogt remembers this period as the high point of his career. He loved working with Marcy and Butler, and was doing historic work with HIRES and the Hamilton, the very instruments he himself had created. For an astronomer, it could never get better than that.

It wasn't just Marcy, Butler and Vogt who had realized that the name of the game was finding smaller and smaller planets. Mayor and Queloz were also racing to the bottom. Shortly after their detection of the first exoplanet, 51 Peg b, the Swiss planet hunters had been asked to build a second version of their groundbreaking ELODIE spectroscope for use on a 3.6-metre telescope that was owned by the European Southern Observatory (ESO). Situated on La Silla, a mountain in the southern part of the Atacama desert, 600 kilometres north of the Chilean capital Santiago, this was one of ESO's prime observing sites.

Mayor and Queloz named the new spectroscope CORALIE and began making detections almost immediately. Their team had grown as they took on more graduate students and collaborators to help them conduct the search. Before the year 2000 was out, they had found 18 more planets. One of them, HD 168746 b, was significant because its mass was probably less than Saturn's. Whereas Jupiter has a mass 318 times larger than the Earth, Saturn's is just 95 times greater than our planet. Crossing this boundary was seen as a milestone, even though there was a long way still to go to get down to an Earth-sized world.

Yet, even as they were working on CORALIE, they were planning its replacement: HARPS, the High Accuracy Radial-velocity Planetary Search spectrometer. Building ELODIE and CORALIE had in some ways been an exercise in compromise. All the way through, the team had been conscious of where they could have done better if more time and money had been available. Now, they were in a position to build their dream instrument because ESO wanted to fund them to build the most accurate spectrometer for planet searching it was possible to build.

Mayor was the principal investigator, Queloz was the project scientist and his opposite number on the engineering side of things was Francesco Pepe, a new recruit at the University of Geneva. Queloz had recently returned to Geneva from a two-year appointment at NASA's Jet Propulsion Laboratory in Pasadena, California. Scientists there had wanted to learn about exoplanet science from him, and he had wanted to learn how a big, well-funded space laboratory worked.

Together, Queloz and Pepe had begun work on HARPS in late 1999, with what Queloz calls the 'dream mission'. He asked Mayor for two weeks observing time on CORALIE. He would take Pepe with him and together they would drive the spectrometer as hard as possible to understand completely where the improvements could be made. Two weeks on a telescope is a large allocation of time, considering how many other astronomers are usually clamouring to make observations, yet it seemed like an investment that could pay off. And it did. When Queloz and Pepe returned, they knew exactly what to do. HARPS was finished by 2003. The instrument's initial accuracy was 1 metre per second and it remains the most accurate planet-hunting instrument in the world.

At NASA, the enthusiasm for exoplanets knew no bounds. Goldin knew that finding Earth's twin was not going to happen overnight. In fact, he banked on it taking a number of years because the search, eventual discovery and subsequent investigation of Earth's twin was an understandable quest that could capture the public's attention and carry the agency forward. He put together a working group of scientists to look at the missions NASA already had on its books, and those that it was planning. He wanted to know how they could help in the search, and he wanted to know what new missions would be needed to take the eventual picture once Earth's twin had been identified.

But first the astronomers had to be clear what they meant by Earth's twin planet. In some ways the definition was simple, simplistic even. Earth's twin would be a rocky Earth-sized world, circling a Sun-like star in an orbit of roughly the same size as Earth's, i.e. with a radius of about 1 astronomical unit (AU). But what did a Sun-like star mean?

At the turn of the 20th century a dedicated group of female astronomers showed that stars are not all the same, and that the Sun is not as common as we might imagine. These women were known as the 'computers' of Harvard College

Observatory, Massachusetts, and they were employed to catalogue stars according to their spectra.

As Wollaston and Fraunhofer had discovered in the early 1800s, the Sun displays dark lines in its spectrum that relate to its chemical composition. The Harvard team set out to catalogue the stars by looking at their spectral lines and swiftly discovered that, although some stars show similar patterns to the Sun, others show subtle or even gross differences.

Working as a team, the computers compiled the first new way of classifying stars since the magnitude system of the ancient Greeks. They were led by Williamina Fleming, a Scottish emigrant who had first been employed as a maid by the observatory's director, Edward Pickering. Legend has it that his male assistants were so lax that he thundered one day, 'Even my maid could do better.'

She certainly could. In 1890, she published the *Draper Catalogue of Stellar Spectra*, named after the Boston doctor whose bequest had made her research possible. It contained 10,351 stars that Fleming had arranged into sixteen categories, labelled A–Q. The groupings depended upon the pattern and strength of the spectral lines. For example, those stars showing strong hydrogen lines were placed into categories A–D.

Another Harvard computer then took up the work. Annie Jump Cannon was a deaf physics teacher with extraordinary powers of concentration. Using better data, she reordered the categories so that the various spectral lines faded in and out across the whole scheme. In doing this, she ended up dropping most of the categories. By 1912, with her team now classifying stellar spectra at a rate of 5,000 per month, she presented the sequence that is still in use today: O, B,

A, F, G, K, M. She even proposed a mnemonic by which to remember it: 'Oh Be A Fine Guy/Girl Kiss Me!'

The sequence is now known to mirror the temperature of the stars. O stars are the hottest at 25,000 degrees Celsius, whereas M stars register just 2,000 degrees Celsius. The Sun is a G-type star with a surface temperature of 6,000 degrees Celsius. Finer gradations became possible with improved instruments, and nowadays the letter classification is followed by a number from 0 to 9. On this scale, the Sun is a G2 star.

Clearly the temperature of a star is a factor in how close a planet could be in order to be habitable. It is not sufficient simply to find an Earth-sized world in an orbit of roughly 1 astronomical unit around an O star or a M star. The first would make the place too hot, the second would make it too cold. It had to be a Sun-like star, and by that astronomers meant G-type at least and G2 if possible.

Astronomers imagined that there could be some small leeway in the orbit. They came to talk about the Goldilocks zone or the habitable zone, where temperatures would allow liquid water to be present in large quantities on an Earth-like planet's surface. Using our solar system as the archetype, they roughly defined the habitable zone as extending from beyond Venus's orbit (0.75 AU) to inside Mars's (1.5 AU). Earth sits more or less in the middle of these.

With these considerations, astronomers could think of Earth's twin as being an Earth-sized world around a G-type star, which forced it to have an orbit of about 1 AU.

With this definition in hand, Goldin's NASA study group went on to find an elegant unifying principle that could be applied to many of the agency's endeavours. It was the search for our cosmic origins. In a document produced by NASA's

Jet Propulsion Laboratory in the 1990s,[11] the Origins Program was described:

> How did we get here? How did stars and galaxies form? Are there other planets like the Earth? Do other planets have conditions suitable for the development of life? Might there be planets around nearby stars where some form of life has taken hold? These questions have intrigued humanity for thousands of years. Astronomers approach these fundamental questions by looking far into the Universe, back toward the beginning of time, to see galaxies forming, or by looking very close to home, searching for planetary systems like our own around nearby stars.

The Origins Program defined two ambitious space missions. One was the Next Generation Space Telescope (NGST), which was going to look into the far distant reaches of the universe to observe how the first stars and galaxies formed. The second was the Terrestrial Planet Finder (TPF), which was envisaged to be a collection of four modest space telescopes, each about a third of the size of the Hubble Space Telescope, mounted in a 75-metre-long frame. Work had already begun on the NGST but the TPF was only at an early stage of discussion.

On paper, the four telescopes of TPF could be used to pick out the faint starlight reflected off an Earth-like planet's atmosphere, allowing the planet to be analysed. Although astronomers and engineers would hesitate to call the resulting data an image, the general public would not recognize the subtlety. The raw data from TPF would look like a small spot of light where the planet was located. It was something that could be pointed to and announced 'this is a planet like Earth'. It would be a monumental achievement, and one that would spur the development of even more ambitious space missions to try to resolve the details of the planet.

All that was needed was engineering skill, which required money, which required politics. NASA is a state-run agency that depends on the American tax payer footing the bill. The NASA Administrator works at the agency's headquarters in Washington DC, just ten minutes by car from the White House. It is the Administrator's job to liaise with government to win the best deal for NASA year by year, and Goldin set to work convincing people about the Origins Program. It was a compelling case because NASA was already working on a space mission that could be rebranded as a precursor to this grand effort. This was where American astronomer Charles Beichman came in.

He had begun his academic studies in 1970 as a philosopher at Harvard College. His original interest in the Presocratic Greek philosophers was fuelled by the big questions that they asked about the origin of the universe, and of planets, and of life. It was during his first year that he had his own eureka moment (without the bath or the nudity) when he realized that modern astronomy was developing the tools and techniques needed to address the questions he was most interested in. These big questions were no longer confined to philosophical investigation; they could now be pursued using the tools of modern science.

He switched courses to astronomy and graduated magna cum laude in 1973. At the time, infrared astronomy was in the ascendancy thanks to new detectors that were coming onto the market, and he naturally gravitated towards space missions that were being developed to exploit this new advance. Infrared consists of longer wavelengths than visible light. It carries energy that we cannot see but which we perceive as heat, and is the predominant emission of objects at room temperature and below. Planets also give off infrared radiation, as do far distant galaxies because of the young stars that are forming inside them.

Beichman had worked his way up the career ladder to become the director of the NASA-funded Infrared Processing and Analysis Center at the California Institute of Technology (CalTech). The position brought with it an association with NASA's Jet Propulsion Laboratory (JPL), in Pasadena, where space scientists and engineers dream up new missions. JPL is managed for NASA by CalTech, so Beichman was the perfect person to steer the science of the Origins Program, in particular, the route to TPF. In 1996, he was appointed Origins Scientist at JPL, where NASA was working on two missions that could immediately be rebranded for the Origins Program.

The first was the Space Infrared Telescope Facility (SIRTF). It was already being built and was seen as an essential stepping stone because both NGST and TPF were designed to work in the infrared region of the spectrum. SIRTF was eventually renamed the Spitzer Space Telescope and was launched in 2003.

A second precursor mission was the Space Interferometry Mission (SIM). Interferometry is a technique pioneered in the late 19th century to combine two beams of light to measure much smaller distances (and other properties) than is otherwise possible. One of the first astronomical uses of an interferometer took place in the second decade of the 20th century at the Mount Wilson Observatory in California.

After a stint in the US Navy during the First World War,[12] Polish émigré and Nobel Prize winner Albert Michelson set about constructing an astronomical interferometer because he realized that it could offer much sharper resolution than ordinary telescopes. This improvement could be used to measure the diameter of stars and resolve the closest binary stars into separate components. To provide the two beams of light, he and his assistant, Francis Pease, placed two small mirrors 120 inches apart on a bar of rigid steel and placed this over the 100-inch-wide telescope. The mirrors were set at 45 degrees and bounced the starlight towards the opposite mirror. Before they could touch, the beams struck two more 45-degree mirrors that directed the starlight down into the telescope where they were brought to a focus. For the interferometer to work, the path lengths of the two separate beams must be absolutely identical.

Working to perfect an interferometer of 120-inch (10-foot) separation, Michelson and Pease then set about building one twice the size. By 1920 they succeeded, and with this larger instrument they measured the diameter of the giant star Betelgeuse. They found it to be 240 million miles across,[13] about the same size as Mars's orbit. Such a great extent was close to the theoretical prediction made by British astrophysicist Arthur Eddington. And with this success the age of astronomical interferometry was born.

As the decades progressed, instead of small mirrors feeding an individual telescope, whole telescopes themselves were increasingly used to capture the light to be combined. When ESO approved four identical telescopes for construction in the Atacama desert of Chile in 1987, they called the project by the singular title of the Very Large Telescope (VLT), and planned that the four 8-metre telescopes could be combined via interferometry. Construction began in 1991 on the mountain top of Cerro Paranal with tunnels being hewn from the rock to house an underground laboratory where the light beams could be brought together.

As well as the four large telescopes, four smaller auxiliary telescopes of 1.8 metres were also constructed to provide extra light to the interferometer. Then began the painstaking work of perfectly aligning all the optics. This took until 2001 by which time, in California, JPL personnel were building similar but smaller equipment in their labs in Pasadena to take the technique into space via SIM.

The mission began with a four-month feasibility study, which led quickly to the award of two contracts worth $200 million. TRW and Lockheed Martin were pitted head-to-head to come up with a viable design that would allow SIM to look for nearby rocky planets. It would not be able to see the planets directly – this would be for the later TPF mission – but it would infer the planet from watching the star wobble. Unlike the radial-velocity searches that looked for the Doppler effect imparted by the wobbling, SIM would actually see the star wobbling for real. It would measure the star's position so accurately that it would see the tiny pirouette as the star was dragged around by the gravity of the planet.

Launch was scheduled for 2005.

SIM was not just an exoplanet mission. There were a number of other astronomical studies that its extreme precision could assist, such as helping to determine distances across the universe, but it was the exoplanet application that NASA really played up. They were especially vocal about the fact that SIM would have the sensitivity to detect planets just a few times the mass of Earth. These were dubbed super-Earths. Considering that the planets being discovered by the radial-velocity astronomers were several to many hundreds of times the mass of Earth, SIM represented a giant leap in detector technology.

Interferometry was another of the reasons NASA became interested in the Keck telescopes. The two identical giants were designed to allow their light to be combined, rather like the VLT, and so this allowed NASA a chance to learn how to do this on Earth before attempting to do it in space. By 2000, the industrial studies were progressing, but more slowly than hoped. The technology needed to perform interferometry in space was proving daunting, and launch was delayed until 2009. Still, NASA felt confident enough to appoint a science panel to the mission, and heading the list was Geoff Marcy.

Although all fields of the mission's science were covered by scientists on the panel, the list of questions that had defined the Origins Program in the beginning had been whittled down to just two: How did we get here? Are we alone?[14] Without explicit mention of the cosmology and other purely astrophysical goals of the mission, some began to feel threatened that the exoplanet programme would start devouring extremely large amounts of cash, leaving them without money to perform their own, more traditional astronomical research. To start with, these seemed like internal worries only. Beyond the professional community, as Goldin had realized, the public were lapping up the search for Earth's twin. Despite being told that TPF was at least a decade and a half away, and that NASA were studying whether they could even build the spacecraft in that timeframe, nothing mattered; the public appetite was voracious. For a while, it was Goldin's dream come true: NASA's bold new purpose. But as forward thinking as the mission appeared to be, if anything NASA was playing catch-up. Across the Atlantic, ESA already had what looked like a workable design, with one man at the helm driving it all forward.

Malcolm Fridlund had originally worked on the PRISMA and STARS mission proposals (see Chapter Four), turning them towards the study of exoplanets, before almost dying of blood poisoning. When he returned to work in 1997, the mission was in the process of being transformed into Eddington by his replacement, Fabio Favata. With no place for him on that project anymore, Fridlund was offered a new study: a mission called Darwin. He was less than enthusiastic at first.

He had been at the agency almost a decade by this point, yet not a single one of the missions he had studied had made it off the drawing board. He had passed up a research position in London to work for ESA because he had been smitten by the possibilities on offer. It is not hard to understand why. The European Space and Technology Research Centre (ESTEC), in Noordwijk, The Netherlands, is an impressive place. Europe's equivalent to NASA's Jet Propulsion Laboratory, it is the size of a large village nestled behind the sand dunes that shelter it from the North Sea. But most importantly, if you are going to be a player in the scientific exploration of space, ESTEC is the place to be. It is often where missions are born, and every ESA spacecraft passes through its test chambers en route to space.

Offered a place at ESA as a research fellow, Fridlund joined at once and, just a year later, he was offered a permanent position as a staff scientist. He was soon put to work on studying what science could be done from the surface of the Moon. This was in the wake of President Bush's 1989 call for a return to the Moon, and Europe did not want to be left behind. Fridlund's task was to study the science that an interferometer on the surface of the Moon could perform. The idea was that ten rocket-propelled telescopes would land and be used together to simulate a much larger telescope, like the fields of radio dishes that dominate some research institutes around the globe.

The study was well received at the time, but found to be too technologically challenging for immediate consideration. So Fridlund had moved on to studying smaller, more feasible missions and was very disappointed when they too were not chosen for development. When he was asked to study Darwin, he immediately saw that it was an extraordinarily ambitious space interferometer requiring a number of identical, formation-flying space telescopes to work. He told his superiors that he wasn't interested in working on another mission that was destined never to make it off the drawing board. They made a bargain with him. If he could show in six months that it was unworkable, they would kill the project stone dead and move him to something more fruitful. So Fridlund set about finding all the faults he could with Darwin to show that it could not possibly be achievable.

The proposal had originally been received in 1993 during an open call for mission ideas from European scientists. It was essentially Terrestrial Planet Finder but had preceded its NASA rival by a good handful of years. Wasting no time in his attack, Fridlund zeroed in on what he thought was the most probable show-stopper: that the mission would have to be sent way out to Jupiter in order to work.

Spend time under a clear, dark sky and sooner or later you will see a shooting star or two. These are minute dust particles burning up in our atmosphere. They belong to a population called the zodiacal dust, which lives in the gaps between the planets of the solar system. It is constantly replenished from the tails of comets and the occasional collision between asteroids in the belt between Mars and Jupiter. Each particle is an efficient radiator of infrared emission, and Fridlund thought that en masse their combined emission would blind the Darwin mission. So the spacecraft would have to be sent out to the orbit of Jupiter where the dust was thinner and so less bright.

This was daunting. ESA had launched a mission in 1990 called Ulysses to study the Sun's polar regions, but to reach there they had to send the probe out to Jupiter and use the planet's large gravitational field to sling-shot the craft out of the equatorial plane and back so that it would pass over the top of the Sun, and then underneath. To do this, Ulysses had been launched from a NASA space shuttle in Earth orbit, using an attached rocket to boost it on a 16-month cruise to Jupiter. It was difficult to imagine how to do this for a reasonable amount of money with the multiple telescopes needed for the Darwin mission. So Fridlund began studying the zodiacal dust. What he found was a surprise: so long as you observed at right angles to the ecliptic, the name for the plane of the solar system, the dust was not a problem. This meant Darwin could operate much closer to Earth and therefore slash the cost of launch to something affordable. Instantly, one of the major objections to the mission crumbled into . . . well, dust.

Fridlund began looking at other aspects of the mission and, after six months of work, came to his conclusion. Going completely against his intuition, he could find no immediate show-stoppers to the Darwin project. Yes, it was hugely ambitious because no one had attempted a mission like this before, and yes there would need to be technology developed specifically for the mission, but both of these things were what the space agency was set up to do. Far from killing the project, he recommended getting industry involved to help define what technology would need to be developed to make it work. In 1997, Alcatel in Cannes, France, won a three-year contract with ESA to develop a mission architecture.

They found that the mission was achievable with a flotilla of eight spacecraft: six would be 1.5-metre telescopes, similar in size to that being planned for the Eddington mission. All would be flown in a hexagonal formation, about 25 metres apart from one another, at a spot 1.5 million kilometres away from Earth (a fry cry from the 2.4 *billion* kilometres originally thought when Jupiter's orbit was the target). Each telescope would beam its light to a central 'hub' spacecraft that would combine the rays and record the information. An eighth spacecraft would sit somewhat 'behind' this constellation and handle communications with Earth. The engineers had found a way to pack all eight spacecraft into the nose cone of an ESA Ariane 5 rocket, one of the most powerful on the planet, so that launch could be achieved in a single go. No previous unmanned mission had ever been this daring or demanding but it was achievable. With hard work, the mission could be ready to fly by 2012.

The basic mission, known as the baseline, was for the spacecraft to operate for five years. The first years would be spent targeting 200 stars, most of them G-type stars similar to the Sun, within a distance of 33 light years. This would be the detection phase, in which Darwin would look to find rocky planets in the habitable zone around the central star and would generate the family portraits of these solar systems. It was no easy task. The problem with seeing a planet like Earth around a star like the Sun was one of contrast. Stars generate their light and shine it into space whereas planets only reflect a star's light. At optical wavelengths, where the star is brightest, the planet will be outshone by a factor of a billion. Seeing it would be like trying to pick out a table-tennis ball held next to a searchlight, while standing a kilometre or two away. Moving into the infrared made the problem somewhat more tractable because stars like the Sun emit less infrared than visible light. Also planets give off some of their own infrared, and so the contrast drops to about a million to one.

It would still be daunting to pick out the planet's light but an electrical engineer from Stanford University, California, had published a theoretical solution back in 1978.[15]

Whereas an interferometer is usually used to reinforce the light from the object it is looking at, Australian American Ronald Bracewell had shown that a simple readjustment in the position of the telescopes would cause the light from the central object to be cancelled. Such a configuration is called nulling interferometry. Any planets in orbit would not be directly in the centre of the interferometer's line of sight and so would not be cancelled, leaving them visible. So Darwin – and TPF in the States – was designed to work as a nulling interferometer.

Once the detection phase of the mission was over, scientists would choose the best 50 planets for follow-up studies. This would involve splitting the planet's light into a spectrum to analyse the composition of the planet's atmosphere. Each planet would have to be observed for weeks at a time to build up the necessary amounts of light to perform the analysis but there was a staggering pay-off to be had for sticking with it. Not only would Darwin be able to isolate the light from an Earth-like exoplanet (in effect take its picture); not only would Darwin be able to look for the major constituents of the exoplanet's atmosphere; but scientists would also be able to tell from those constituents whether there was evidence for life on the planet because, in the very act of breathing, we affect the composition of our atmosphere. In short, Darwin was nothing less than a scientifically valid method of searching for extraterrestrial life.

Searching for life in the universe had always been somewhat outside the realms of respectability. The Search for Extraterrestrial Intelligence (SETI) has a chequered history and is regularly criticized, even by other astronomers, as being too speculative to waste money on. This is partly because building a set of radio telescopes capable of eavesdropping on stray communications from a distance of hundreds of light years would cost more than the Apollo Moon-landing programme, so more affordable efforts had to rely on the assumption that alien civilizations would be blasting powerful radio beacons into space purposely to attract our attention. Such bold assumptions do not sit well with many scientists. A NASA SETI programme launched with great fanfare on Columbus Day 1992 was shut down less than a year later at the command of Congress.

Critics have often pointed out that on Earth it has taken 4.5 billion years from the appearance of the first microbial life on our planet to reach the state of intelligence necessary to build radio transmitters. If this is the pattern for all planets then, unless technological civilizations are extremely long-lived, there could be a highly significant number of planets on which life (even intelligent life) is present but that are not broadcasting radio. The scepticism that surrounds SETI has meant that no one but a few 'conviction scientists' have wanted to attach their colours to the mast. However, Darwin was different.

It did not rely on the presence of intelligent life to generate signals, it simply relied on the presence of life. This could be in the form of intelligent creatures, dumb animals, or pond life. All that mattered was that they were breathing (or taking in some form of food and metabolizing it to extract energy) because in doing so they would change the chemical composition of the atmosphere.

With no life forms present on a planet, the composition of the atmosphere is determined solely by chemistry. As soon as life begins, the rules change: biology is also helping to determine the composition of the atmosphere.

Take the oxygen in Earth's atmosphere for example. This was originally generated by cyanobacteria, more than 2.3 billion years ago. The cyanobacteria evolved to generate their energy through photosynthesis, the same way plants do today. In the process of metabolizing the sunlight, they gave off oxygen as a waste product. Although it was initially absorbed by rocks and the oceans, the oxygen saturated these reservoirs in a few hundred million years and then began to fill the air.

The Great Oxygenation Event, as it is known, led to the evolution of micro-organisms that could use the gas to fuel their metabolisms. It also led to a widespread extinction of existing organisms, which found the oxygen toxic. Some researchers even call it the oxygen catastrophe but from our point of view it was a good thing because the chain of evolution eventually led to humans and our oxygen-hungry brains.

The bottom line is that without life there would be significantly less oxygen in Earth's atmosphere. If all life disappeared tomorrow, the oxygen would gradually be absorbed back into the rocks. Even now, most of the oxygen to be found on Earth is bonded into the minerals inside rocks. So looking for the presence of oxygen in an exoplanet's atmosphere by splitting the infrared light into a spectrum with the Darwin mission would be a good first attempt at assessing whether a world is an abode of life.

As luck would have it, the spectral absorption line signatures (see Chapter One) of water, ozone and carbon dioxide can all be found next to one another between wavelengths of 6 and 18 microns, the very range of infrared wavelengths where the star–planet contrast drops to a manageable million-to-one. This was a major boon because the carbon dioxide would show that the planet had an atmosphere, the ozone would show that there was oxygen in the atmosphere, and the water would indicate that there were probably seas or oceans on the planet. In our own solar system, whereas Venus and Mars both show carbon dioxide, only Earth shows water and ozone too. Scientists all agreed that in the search for an Earth-sized world these gases would be a very strong indication that it was a living planet. The life may not be intelligent but it would be life.

Also on the plus side, building infrared space telescopes made sense to ESA. It was in the throes of constructing Herschel, a 3.5-metre infrared telescope that became the largest telescope to fly in space upon its launch in March 2009.[16] It remains the largest to this day and will continue to be so until the renamed NGST (now known as the James Webb Space Telescope: JWST) is launched in 2018 or later.

On the downside, although Bracewell had developed the mathematics of a nulling interferometer, no one had ever built one on the ground, let alone in space. To bridge this gap, Fridlund proposed a small space-based mission to test the technology using tin-can-sized telescopes. The same knowledge gap had been perceived in America also. This was another reason for NASA buying into the Keck observatories, so that they could perform nulling interferometry and convince themselves that it would work.

As studies on both sides of the Atlantic progressed, it became clear that they should join forces. A mutual strategy to develop the technology and the techniques would allow them to share the costs> It would also allow them to share expertise, and eventually it would allow them to construct and run the mission jointly. Such ESA–NASA collaborations had been pioneered on solar missions and shown to work well for all concerned. Fridlund thought it made more than just economic sense. 'As well as the benefits of not having to develop all the technology and spend all the money yourself, this was such an important mission for mankind that we thought it was right to do it together,' he remembers.

In the early years of the new millennium, everything was superb. ESA and NASA were like relay runners, each taking up the baton when the other finished a stretch. As European teams solved one problem, their American equivalents would solve another. It led to a time of enormous optimism and excitement. More help came from the industrial teams. Unusually, they were investing their own money in the project, rather than relying solely on grants from ESA. Spurred by their belief in the goal, the industrial partners were hiring staff and funding graduate students to work on the underlying aspects of the endeavour. Everyone was driven by the chance to reach this fundamental milestone in human history, to find Earth's twin and see if it cradled life.

But not all at ESA was good news. Just down the corridor from Fridlund's office in ESTEC was that of Fabio Favata, the Eddington project scientist. This was the agency's mission – to find transiting exoplanets – and Favata was fighting for its life.

SIX

Strange new worlds

It was 11 December 2002. No one working on Eddington at the time had any reason to suspect that the mission was in danger. But it was. Thousands of miles away in French Guiana, sitting on a tropical launch pad, Ariane flight 157 was poised for a night launch. The 53-metre-high white giant sat wreathed in vapours as the countdown reached zero. This was the maiden voyage of a more powerful motor that would allow the rocket to lift 10 tonnes into geostationary orbit, where most of the communications satellites reside. To test its new heft, it was to carry two such spacecraft on this flight. As the clock reached zero, the main rocket engine and the boosters ignited, and it took to the sky.

When it came, the disaster was not a particularly spectacular one. Cameras on the ground had almost lost sight of the rocket. They just managed to capture the booster rockets being jettisoned, and to the eye everything looked normal. But on the computer screens in mission control, it was clear that something was going wrong.

Flight controllers could see that the cooling circuits on the main engine were in trouble. This led to the partial breakup of the main engine nozzle, which meant that the rocket started to lose control of its trajectory. When the nose cone hiding the payloads at the front of the spacecraft opened and separated as planned, the rocket lost even more control, and by the time it had veered 150 kilometres off course an automated self-destruct system cut in. The rocket exploded, scattering debris into the Atlantic Ocean.

Apart from the people operating the rocket and the satellite owners, one other person immediately knew what a catastrophe this was. He was David Southwood, Director of Space Science at the European Space Agency. He had been keenly watching the launch because he was planning to use exactly the same configuration to launch a science mission called Rosetta in a month's time.

Now widely known to the public because of the audacious touchdown of its Philae lander on the comet 67P/Churyumov–Gerasimenko on 12 November 2014, back then the mission had not even left the ground and was planned to visit a completely different comet. It was Southwood's responsibility to get it safely on its way. Were the same to happen to Rosetta on 12 January 2003, the world would watch helplessly as one billion euros of tax payers' money fell to the bottom of the ocean.

Such a failure was unthinkable. The science programme at ESA had been on the up, truly establishing itself on a par with NASA, with missions that ran the entire gamut of astronomical and cosmological science. As a result of this new-found confidence, Southwood had been presiding over an ambitious set of endeavours.

As an enquiry got underway, which would eventually reveal that the problem lay in the cooling circuits, Southwood found himself with a very difficult choice to make: gamble that everything would be fine, or delay the launch and lose the target comet for the mission. In the event, there was no choice to make. The inquiry board swiftly concluded that the rocket would not be ready to fly again until the second half of 2003, giving plenty of time for the engine problems to be fixed, tested and verified. Rosetta was grounded.

The mission may have been saved but there was a price to pay, literally and metaphorically. The spacecraft now needed to be stored, parts of it would need to be refurbished because of the corrosive nature of the rocket fuel used, and the science teams would have to go back to the drawing board to find a new target. All of this would cost money but every euro cent of ESA's Science budget was already spoken for. Where was it going to come from?

There were other missions too that were soaking up more money than originally foreseen. ESA's infrared space observatory Herschel was proving more difficult to build than envisaged. By June 2003, the cash-flow problem was becoming so serious that it was being debated by ESA's advisory council. Southwood successfully lobbied them for a loan of €100 million to get him over the hump but this was to be paid back from the science budget by the end of 2006.

Planners began to think where the cuts could come from, and how some level of contingency could be built into the budget to prevent such a crisis occurring again. Money could be saved by delaying the start of some missions, others could be 'descoped' and made simpler, but it just was not enough. The only way to repay the loan, fulfil the majority of ESA commitments and build in some contingency was to axe a mission. Two missions stuck out like sore thumbs: one was a technology demonstrator called LISA-Pathfinder, the other was Eddington. The two would be pitted against each other, and the respective mission teams began to hone their arguments for why they should be retained.

LISA-Pathfinder was a mission that would prove the technologies required for a new breed of spacecraft that could investigate gravity across the universe. At the time, there was an agreement in place between ESA and NASA to use the experience gained on LISA-Pathfinder to build an extraordinary mission called LISA (Laser Interferometer Space Antenna). This was defined to be capable of measuring the minuscule ripples known as gravitational waves. Cancel LISA-Pathfinder and this transatlantic collaboration was placed in jeopardy.

Put like that, Eddington didn't really stand a chance. After a final round of presentations at ESA's Paris headquarters in front to the Science Programme Committee, the painful decision was made to cancel Eddington altogether. It was a horrid blow. At the same time, the plan to land on the inner planet Mercury was scrapped although an orbiting spacecraft was retained.

Ironically, the Americans subsequently pulled out of the LISA collaboration, and many other ESA–NASA agreements, in the wake of the 2008 economic crisis. Undaunted, ESA pressed ahead with LISA-Pathfinder, which launched in 2015, and now plan to build some version of the LISA mission on its own for launch around 2028.

There was some comfort for the Eddington scientists. To highlight the fact that there was never any doubt about the science goals of the mission, ESA diverted some money to a smaller French mission called CoRoT (Convection, Rotation and planetary Transits). With a mirror just 27 centimetres across, it had originally been a precursor of sorts to the five-times-larger Eddington spacecraft and a number of CoRoT scientists had also been members of the Eddington collaboration. ESA provided the telescope optics, computer processing units and a baffle to keep out stray light and make it as precise as possible. It then tested the spacecraft's payload at ESTEC.

In America, NASA were building the 1.4-metre Kepler space telescope and everyone knew that once launched the NASA publicity machine would drown out everything – and with some good reason. Kepler would be much more sensitive to planets than CoRoT. Whereas the French mission was expected to see planets a few times larger than Earth and above, the American spacecraft was designed to see Earth analogues. So the smaller French-led mission had to launch on time, giving it a few years in the limelight before Kepler's discoveries blew it away.

Nevertheless, the loss of Eddington was a terrible blow to European exoplanet hunters. Unfortunately it was just the first of many more bruises to come. No one could have known just how bad it was going to get, especially because, as far as the original exoplanet hunters were concerned, these were the salad days. Marcy, Butler and Vogt in the US, Mayor and Queloz in Europe, were announcing discoveries on a regular basis. Each newly discovered planet seemed stranger than the last, all inexorably leading down the path towards Earth's twin.

As more telescope time became available for exoplanet programmes, the astronomers began to widen their search. Initially they had targeted just the G-type – Sun-like – stars but these are by no means the most numerous stars in the Galaxy. They make up just 10 per cent of the Milky Way's population. By far the more numerous stars are smaller, dimmer examples called red dwarfs. On the Harvard classification scheme they are listed as K- and M-type stars. Being more numerous, there are naturally more of them closer to Earth, and so they make attractive targets for the exoplanet hunters. As luck would have it, the definitive catalogue of these nearby stars was published in 1993. It was the life's work of a German astronomer called Wilhelm Gliese.

As befits an astronomer, he was born on the summer solstice, 21 June 1915. His interest in the celestial realms was piqued at eight when his grandmother gave him a calendar that featured beautiful drawings of the Moon's phases. By 13, he was regularly observing the night sky to chart the brightness fluctuations of certain stars.

When the Second World War came, Gliese had just finished his Ph.D. studies; he was awarded his doctorate in 1942 while serving in the army on the Eastern Front. His scientific studies were further interrupted when he was captured by the Soviet army in the final year of the war and, although he was not repatriated until 1949, he made the most of his imprisonment by giving scientific lectures to his fellow inmates and making brightness estimates of the stars with his unaided eyes. He smuggled home these observations in notes hidden inside cigarette packets and published his results in 1950.

Once safely back in Germany, and working for the Astronomy Computing Institute of Berlin-Dahlem, he began to compile star charts. His interest in nearby stars began at the suggestion of Dutch astronomer Peter van de Kamp, who was to gain notoriety for his erroneous belief that he had found planets around the nearby red dwarf called Barnard's Star (see Chapter Two). At the time, just 40 stars were known to be particularly close, and van de Kamp wanted more targets for his planet search. Beginning the work, Gliese soon became gripped by the challenge of finding nearby stars. Poring through reams of data, he published his first catalogue in 1957. It contained almost one thousand stars. He updated it in 1969 and 1991, by which time the catalogue had grown to some 3,800 stars, all within 81 light years of Earth. The final edition was published in 1993, the year after his death. It furnished the exoplanet teams with good targets, and the searches were really bearing fruit.

By the summer of 2004, almost a decade had passed since the discovery of the first exoplanet, 51 Pegasi b, and more than 120 had been added to the roster, most of them hot Jupiters. This was important because, as exciting as those first exoplanets had been, most astronomers had thought they were oddballs that had shown up in the data only because they produced the largest wobble in their host star, hence were the easiest to see. As the years passed, however, what was becoming clear was that the hot Jupiters were plentiful and formed a distinct class of planet in their own right. As such they could not be dismissed as freaks; clearly there was a pathway in nature that commonly led to their formation, and the theorists were playing catch-up to dream up what that might be.

They had been modifying their computer simulations, or models, to include the effects of gravitational interactions between planets and the discs of material from which they formed. The results were a revelation.

At first a planet-forming disc is made of gas with about 10 per cent dust mixed in. As Kepler had described in his third law of planetary motion, the further an object is from its central star, the slower it will orbit. This applies to the gas and dust just as surely as it will to the eventual planets. Because the gas and dust are distributed continuously, there will be friction between the orbits as the faster-moving inner material 'rubs' against the outer slower layers. The friction will heat the matter, causing it to radiate away energy. This release of energy causes the dust and gas to gradually spiral in towards the central star. So there is a natural tendency for the material to migrate inwards.

Next, the simulations showed that, as a planet begins to form, its gravity will affect the distribution of matter around it. Dense rivers of gas will appear and flow into the planet from either side. These will have some gravity of their own and so will affect the orbit of the planet, acting like an anchor to slow it down. As a result, the planet will gradually spiral inwards towards the star. This is known as type I migration. Computer models show that it occurs more rapidly the more mass a planet has accumulated. However, once a planet has reached around 10 times the Earth's mass, its gravity is enough to clear away most of the matter in its immediate orbit. Gas still enters the gap from the outer edges of the disc, where it is consumed by the planet, and so the process of migration continues, although at a slower rate than before. This is known as type II migration and is thought to be responsible for the hot Jupiters.

It was also realized that the more eccentric planets, those on highly elliptical orbits such as the third discovered exoplanet, 70 Virginis, could have been gravitationally scattered. This occurs when two planets of similar sizes have a near-collision. One will have the overall size of its orbit increased; the other's orbit will be decreased. The planet which has its orbit increased could even be given enough energy to escape its star altogether and wander interstellar space. Significantly, the shapes of the orbits will be changed too. It is highly unlikely that a planet will remain in a circular orbit after a gravitational interaction. So the more eccentric an exoplanet's orbit, the more likely it is to have been the product of a near-collision between forming worlds.

At first the success of these migration scenarios were a source of comfort because we at last understood how these impossible planets could form, but then they became a worry. The simulations were showing that migration was easy – and deadly.

The problem was that the wandering gas giants would have pushed any interior planets into the central star. If our own Jupiter had wandered like this it would have pushed Mercury, Venus, Earth and Mars to a fiery fate inside the Sun. Corroborating this growing realization, the exoplanet hunters were beginning to notice that the stars hosting hot Jupiters all displayed higher quantities than average of planet-building chemicals in their atmosphere. Was this the debris of incinerated planets? This was a worry because, if migration was easy, it could mean that our solar system is the freak. If so, finding Earth's twin might be a nearly impossible task.

This concern fuelled the first reservations about ESA's Darwin mission and NASA's Terrestrial Planet Finder, both of which were designed under the assumption that our solar system's arrangement of planets was widespread around other stars. Addressing this concern added urgency to the Kepler mission, which was being built at the time. Designed to monitor hundreds of thousands of stars continuously for years, it was intended to find the frequency of Earth analogues out there. But until it launched at the end of the 2010s, the onus was on the radial-velocity teams to find smaller and smaller planets, and multiple planet systems. In August 2004, they crossed a milestone.

Back in 1997, Butler and Marcy had announced the discovery of a hot Jupiter around the nearby star 55 Cancri A. Located just 40 light years away, 55 Cancri A is a Sun-like star that is part of a binary system. It is orbited by a red dwarf star (55 Cancri b) at a distance of more than 1,000 astronomical units (AU).[17] Exoplanets are labelled with lower case letters to differentiate them from stars: for example 51 Pegasi b shows that this was the first exoplanet to be discovered around the star 51 Pegasi. The hot Jupiter found around 55 Cancri A should strictly be termed 55 Cancri Ab, however, most astronomers colloquially refer to it as just 55 Cancri b. It has a mass at least 78 per cent that of Jupiter and an orbit of 14.7 days. This places it just 0.11 AU from its parent star, yet, as certain as the astronomers were of this planet, its presence could not explain all of their data. Once they subtracted the wobble it induced, the star should have appeared to stand still but it didn't; there was more movement. It could only mean that another planet was pulling on the star.

The team made more intense observations. They were still using their original set-up at the Lick Observatory and by 2002 had extracted two more planetary signals from the data. The first of these, 55 Cancri c, had an apparent orbit of 0.24 AU, an orbital period of 44.3 days and a mass at least 0.1 of Jupiter. The second, 55 Cancri d, was much larger, at least 4 times the mass of Jupiter, but it travelled in an orbit that the researchers knew would raise eyebrows. They highlighted it in the title of their paper: 'A Planet at 5 AU around

55 Cancri'.[18] It strikes a chord with astronomers everywhere because 5 AU is the same distance from its star as Jupiter is from the Sun in our solar system. Yet, even taking these two new discoveries into account, still not all of the radial velocity was fully explained.

To crack the problem once and for all, Marcy and Butler pooled their resources with the Swiss team and a group of astronomers from the University of Texas who had observing time on a large telescope in their home state. They also pulled records out of the Hubble Space Telescope archive. The planet they came up with to explain the rest of the data was nothing like the others in the system. Indeed it was nothing like any exoplanet that had been detected at the time. Far from being a gas giant, it was a comparative minnow. 55 Cancri e was less than 0.06 Jupiter masses, making it just 17 times the mass of the Earth.[19] This is very close to the mass of Neptune in our solar system.

It was not the only smaller planets that the teams were finding. That same month, Marcy and Butler discovered Gliese 436 b. Marcy was heading the investigation and remembers the moment vividly, placing it in the top three moments of his career. He says that his eyes 'bugged out' when he realized that he was writing down figures in Earth masses rather than Jupiter masses. It represented a new era for the discipline. Clearly their analysis skills had crossed a line, giving them the ability to pick out 'small' planets from their data.

At 22 Earth masses, Gliese 436 b orbits its parent star once every 2.6 days. Just 34 light years away, the star and planet were natural targets for follow-up observations and, in 2007, the planet's silhouette was seen to transit the star. It was just a grazing pass across the bright face but it was enough for astronomers to measure the planet's diameter at 4.3 times that of Earth, making it just a bit larger than Neptune, and this gave the average density as 1.51 grams per cubic centimetre, which is just below that of Neptune. With a bit of give and take, everything pointed to this world being an 'ice giant', as Uranus and Neptune are sometimes called. Ice giants are made of rocky cores with a deep gaseous atmosphere but lack the overwhelming amount of hydrogen and helium that dominate Jupiter and Saturn's composition.

With its tiny orbital radius, Gliese 436 b had been assumed to be another migrating planet. Some astronomers speculated that perhaps it had started life as a Jupiter-sized world but had been eroded by powerful stellar activity from its host star. Certainly, its proximity to its star means that it is receiving a roasting. Calculations showed that its daylight hemisphere is likely to reach about 440 degrees Celsius, around twice the temperature of a kitchen oven.

As exciting as this new era of hot Neptunes was, much better was to follow.

Thanks to some updated detectors being fitted to Vogt's HIRES instrument (see Chapter Five) on the Keck, and some newly finessed computer analysis software, Marcy and Butler were able to measure stellar motions to an accuracy of 1 metre per second. As a result, Gliese 876 d appeared from the data like a ship arriving at a misty harbour. It was June 2005. The planet was the third to be discovered around Gliese 876 but was completely different from the other two, or indeed any of the 150 planets discovered around other stars previously. Although larger than Earth, Gliese 876 d was distinctly smaller than the gas giants that had dominated the manifest so far.

It was even smaller than the hot Neptunes that had seemed so astounding just 12 months before.

Calculations showed that Gliese 876 d was at least 5.89 Earth masses. It orbited its star every 1.9 days yet transit searches failed to detect the world. The astronomers estimated that the planet's orbital plane must be inclined around 40–50 degrees away from our line of sight and so the planet would pass above and below the star from our point of view. Although this robbed them of a precise mass, radius and density, they calculated that the most probable mass was 7.5 times that of the Earth,[20] and they felt bold enough to state that it was the first detected planet that stood a good chance of having a rocky surface. They dubbed it a super-Earth, and announced it with a provocatively titled press release: 'Astronomers discover most Earth-like extrasolar planet yet.'[21]

In the release, Butler says, 'This is the smallest extrasolar planet yet detected and the first of a new class of rocky terrestrial planets. It's like Earth's bigger cousin.'

There is no equivalent world in our own solar system; here Earth is the largest of the rocky planets. Marcy was quoted in the press release too, saying, 'This planet answers an ancient question. Over 2,000 years ago, the Greek philosophers Aristotle and Epicurus argued about whether there were other Earth-like planets. Now, for the first time, we have evidence for a rocky planet around a normal star.'

It was heady stuff, accurate and inspiring, and unsurprisingly it captured headlines. Although the astronomers were careful to point out that Gliese 876 d was far too hot for liquid water, at around 200 degrees Celsius, it seemed that they could not resist stoking the fires of expectation with a further quote from Butler saying, 'We are pushing a whole new regime at

Keck to achieve one meter per second precision, triple our old precision, that should also allow us to see Earth-mass planets around sun-like stars within the next few years.'

To add to the excitement, Gliese 876 d was just 15 light years away, making it the closest planetary system to our own that had been discovered at that time. However, more work soon revealed how alien Gliese 876 d might be. Although it was probably a rocky world, it was utterly unlike Earth. It was probably more like Io, the Promethean moon of Jupiter.

Io was one of the original four moons discovered around Jupiter by Galileo in 1610. Invisible to the naked eye, it was revealed by the telescope Galileo had made and sparked a debate among philosophers and theologians about the purpose of these worlds. Later in the century, Isaac Newton suggested that they were spare worlds that God would pick up and move into position closer to the Sun if the Earth ever became uninhabitable.

The true mystery of Io, however, began in 1979 when NASA's Voyager 1 spacecraft flew past. It revealed that Io's surface was devoid of craters. This was a surprise to most planetary scientists, who had expected a barren world similar to our own Moon. The ancient surface of the Moon displays the impact scars accumulated over the last four and a half billion years but Io, which is similarly old, does not. This suggested that some active geological process was taking place that had erased the craters.

Exactly what that process was became clear three days later on 8 March 1979. Voyager 1 was leaving Jupiter and its moons behind after its all too brief fly-by. The navigation team commanded the spacecraft to turn back and take a series of images of the moons so that their position could be measured against the stars. This would allow the operators to calculate the exact position of the spacecraft. The images themselves would be of little scientific value, so they thought, because the moons would be seen in silhouette, backlit by the Sun. How wrong they were.

Navigation engineer Linda Morabito was using a computer to enhance the faint stars on the images. When she was working on the Io picture, she saw a strange cloud on the edge of the moon. This was a 300-kilometre wide fan of lava, shot from the surface by a volcano. The discovery was confirmed by Voyager 2, which flew through the system a few months later in July 1979. These were the first active volcanoes to be found beyond the Earth. The lava constantly spewing onto the surface had erased the craters. As more investigations took place it became clear that Io was a volcanic cauldron. It was never in a state of anything other than catastrophic eruption. Naturally enough, scientists soon asked what was causing this extraordinary volcanic activity.

On Earth, the energy to drive volcanoes comes from the heat released by decaying radioactive elements in the planet's interior. This could not be the case on Io; the smaller quantity of radioactives would have long since lost their potency. Instead it was another Earthly phenomenon that provided the answer: the tides.

A tide is produced by gravity. On Earth, we tend to think of it as the ebb and flow of the ocean but this is just because the water is freer to respond to the Moon's gravity than the rocks. A tidal force occurs because gravity weakens with distance. So the Moon pulls harder on the face of the Earth turned to it than it does on the face of the Earth that is turned away because roughly 12,000 kilometres (the diameter of the Earth) separate these two surfaces. The nearest face is pulled harder than the furthest and the planet elongates a little. Although the majority of this movement is experienced by water, with high tide occurring twice a day, the Earth's rocks rise too but only by about 20 centimetres. This elongation makes it slightly more difficult for our world to spin and so the day is getting progressively longer by a millisecond or two every century.

Of course, the Earth is doing the same to the Moon and, because the Earth is larger, the tide raised on the Moon is greater. This has made it so tough for the Moon to turn that over the course of billions of years, our nearest satellite has lost most of the innate rotation it once had and now keeps the same face turned towards us. This is called tidal locking. The same will happen between a star and an exoplanet if the pair are sufficiently close together.

Io is tidally locked to show the same face to Jupiter but that is not all. At Io the tidal forces are extreme. Jupiter pulls the moon's surface up by 100 metres, the equivalent of a 30-storey building. There are the other large moons in orbit around Jupiter too. Europa, Ganymede and Callisto all pull on Io. As the moons make their way around their orbits, these extra forces sometimes reinforce each other and sometime act against each other. So the tidal force on Io varies constantly, meaning that the tiny world is constantly being squeezed and relaxed, squeezed and relaxed. This constant flux melts Io's interior, driving the volcanoes.

The same process must be at work on Gliese 876 d, only instead of a planet and moons producing the tides, it is the central star and the other planets. As a result, the generated heat will be much greater and a visit to the planet's surface is likely to be a trip to hell.

It seems likely that volcanoes dominate this landscape, driven into a state of continual eruption. The volcanoes themselves will be unEarthly-looking affairs. Being bigger than Earth, the planet's gravitational pull will be exaggerated and so the volcanoes, for all their violence, will probably not be towering affairs. Their rocks will be crushed, turning their precipices into shallower inclines. Yet the greater pull of gravity will ensure that the lava flows quickly down these slopes – faster than even an Olympic sprinter could outrun.

As well as molten rocks, the volcanoes will belch choking gases such as carbon dioxide into the atmosphere. With no plants to feed from this vapour, it will build up and bolster a runaway greenhouse effect, as has happened on Venus in our own solar system. The unbearable heat will make the landscape writhe in the alien equivalent of heat haze. Mirages could make phantom lakes of shimmering water appear to come and go.

If the sky is clear, the parent star will dominate like a bloated orange fruit. It will appear some sixteen times larger than the Sun looks in our sky because, although it is only one-third of the Sun's diameter, it is just 3 million kilometres away from the planet's surface, whereas Earth is 150 million kilometres from the Sun.

The tidal force will be leading to tidal locking too. This process may not yet be complete because Gliese 876 is just several hundred million years old, so the star may not yet have sapped enough rotational energy for the planet always to present the same face. If the gradual process of tidal locking is not quite complete, Gliese 876 d may continue to rotate at a snail's pace. Dawn could take weeks to accomplish, from the first blood-red rays of light appearing over the horizon to the swollen star being fully lofted into the sky. As a result, a day on Gliese 876 d could stretch into months or even years as measured from Earth, whereas its year is just 1.9 Earth days. Were it not for the volcanic activity, the long nights could see the whole night-side hemisphere of the plant freeze over. Without doubt it is a gloriously exotic world. But the revelations did not stop there. It was not long before Gliese 876 d was joined by another similar discovery.

Cursed with an even more outlandish classification, OGLE-2005-BLG-390L b was announced early in 2006. It was found by the same team of Polish astronomers that had detected the second transiting planet back in 2002 (see Chapter Four). This time instead of a dimming they detected a tiny brightening in the star's light caused by gravitational microlensing. The rise in brightness had a small peak superimposed on a larger, broader peak, indicating two bodies in orbit around each other, one larger than the other.

Calculating the sizes of these two bodies showed that it must be a super-Earth in orbit around a red dwarf star but that was as far as the similarity with Gliese 876 d went. If Gliese 876 d was fire, the new world was ice.

OGLE-2005-BLG-390L b is tremendously far away from Earth. Its distance of 25,000 light years places it close to the galactic centre. At five times more massive than the Earth, it is smaller than Gliese 876 d, and instead of furiously orbiting its star in just 2 days, it is calculated to take a leisurely 10 years to orbit once. This places it much further from its star – about three times further than Earth is from the Sun – and astronomers realized that the world was likely to be a frozen wasteland. Its parent star is even more feeble than Gliese 876.

It contains just one-fifth of the Sun's mass and that means it gives out only about one-thousandth of the Sun's light. All told, the planet's surface temperature is likely to hover around −220 degrees Celsius.

Even if there were once oceans on this world, they would now be mostly ice. Indeed, the whole atmosphere may have frozen to the surface because the temperature is just below the freezing points of oxygen and nitrogen, and far below the −78 degrees Celsius freezing point of carbon dioxide. If so, this would make it more like a giant version of Pluto, which hovers around the same surface temperature. The vast majority of Pluto's atmosphere is frozen to its surface, creating an ice-scape that NASA's New Horizons spacecraft photographed for the first time in 2015 during a fly-by.

Under the crushing gravity of OGLE-2005-BLG-390L b, the world would be as smooth as a billiard ball with few, if any, surface features. Any mountains and valleys, plateaux and craters, that did exist would be hidden beneath kilometres of ice.

It is possible that the planet is generating some internal heat, and this may keep pockets of water in a liquid state to form underground lakes. In our own solar system, Jupiter's moon Europa behaves in this way. The surface is covered in ice, yet as spacecraft fly by they detect the unmistakable signature of magnetic activity that could be generated by a layer of water some 100 kilometres deep covering the whole moon. If true, there is 2–3 times more water on Europa than on our own planet.

Earth too provides an analogy for OGLE-2005-BLG-390L b. Deep below the frozen wastes of Antarctica, there is a system of underground lakes and valleys. The most famous of these is Lake Vostok, which has been sealed beneath the ice for more than 14 million years. In May 2012, Russian scientists succeeded in drilling through almost 4 kilometres of ice to reach the lake. Although their preliminary analysis seemed to show nothing, more careful work is now suggesting that there is life in Lake Vostok. If so, it is descended of ancient microbes, trapped there since before the last ice age.

Could OGLE-2005-BLG-390Lb have a similar underground lake system and maybe even life? We will probably never know because the planet is so far away. Astronomers have only glimpsed it once, and the chance alignment that had allowed the microlensing to happen is unlikely to be repeated. Yet this was not a problem because, by now, exoplanet discoveries were coming thick and fast.

This was a bright time indeed for the exoplanet hunters but it wasn't destined to last. In America, particularly, strains were beginning to show.

SEVEN

Divide and be conquered

If the Jet Propulsion Laboratory had been a military organization, you would have called it a mutiny. Scientists and engineers were breaking into factions and beginning to fight over the best design for the Origins' flagship mission, the Terrestrial Planet Finder.

Times were changing at NASA. In 2001, George W. Bush came to the White House as President of the United States and from that moment Daniel Goldin's days as NASA's Chief Administrator were numbered. He had held the post for nine years, serving all the way through Bill Clinton's presidency, and it was entirely within the expectations of tradition that the new President would want his own person in the job.

Goldin had been coming in for some criticism about his realignment of the agency. He had pioneered an approach called 'faster, cheaper, better' as a way of reining in the costly overspends that had seemed part and parcel of the space programme but the strategy was proving only partly successful. Engineers would quip about 'faster, cheaper, better', saying that you can have any two of them but not all three. Bearing witness to their point, three Mars missions – the Mars Polar Lander, Deep Space 2 and the Mars Climate Orbiter, all failed.

Mars Polar Lander and Deep Space 2 failed in their landing attempts and were never heard from again. Mars Climate Orbiter burnt up in the red planet's atmosphere because of an embarrassing error in which NASA subcontractor Lockheed Martin wrote software that worked in imperial units instead of the NASA standard of metric, so all commands sent to the spacecraft to control the thruster were wrong.

The one place where corners could not be cut was when astronauts were involved. There had been 57 space shuttle launches under Goldin's leadership, all without serious incident, and Goldin had steered the International Space Station through launch and into permanent occupation. In doing so, however, the space station programme was $4.5 billion over budget. Clearly, changes had to be made.

Goldin resigned earlier that expected, just a month after the 9/11 terrorist attacks. In his speech, he cited the atrocities as being a wake-up call that he was not spending enough time with his family. Perhaps foreshadowing what was to come, he didn't really mention the Origins Program. Instead, he chose to highlight human Mars exploration, lamenting that his life 'won't be complete until America lands an astronaut on the surface of Mars'.

Perhaps this was a genuine oversight, or perhaps it was because he knew that the effort to design a workable Terrestrial Planet Finder was starting to run into chaos. A small but growing number of scientists believed that, for all the progress that was being made, fundamental hurdles existed in creating the technology needed to make TPF a reality. Chief among their concerns was flying the four or more telescopes and other spacecraft in formation. For an interferometer to work, the beams must be brought together with an accuracy of about a thousandth of the wavelength being combined. So, the shorter the wavelength of the light, the more precisely the distance between the telescopes must be known. At the infrared region that would give the biomarkers of water, ozone and carbon dioxide, the wavelengths are 6–18 millionths of a metre. Many scientists believed that meant flying the spacecraft in formation to an accuracy of a few billionths of a metre. This is an impossibility, and it became a major stumbling block for the mission's credibility.

The concern was a complete misunderstanding as far as Malcolm Fridlund was concerned. Under his leadership, ESA had looked at this problem and concluded that it could fly the telescopes to the same precision that was achieved by spacecraft docking with the space station – just less than a centimetre – but it could measure the positions of the telescopes from each other to nanometre accuracy using lasers. Corrections could then be made in the beam-combiner spacecraft.

But the damage had been done. Frightened by their perception that the necessary formation flying was impossible, the NASA hierarchy began to investigate other ways of finding Earth's twin. In the first four years of the new millennium, they handed out 40 grants to academic and industrial organizations to stimulate thought. Charles Beichman was in charge of this endeavour. He likened it to digging in a field and hitting hard ground, so they would then go and dig somewhere else to see if there was an easier path instead of persevering to get through the tough layer.

Of all the rival designs they studied, one looked promising. It was a device called a coronagraph. These had been developed to block out the blinding light from the Sun's surface so that astronomers could study the much fainter atmosphere, which is called the corona and gave the instrument its name. They were increasingly being used to study other stars. For example, it had been a coronagraph that had revealed the dusty disc of material around Beta Pictoris. So a breakaway group at the Jet Propulsion Laboratory began studying whether a single space telescope using a coronagraph could be used to block out a central star's light in order to detect any Earth-like planet in orbit around it.

TPF-C, as the design became known, was to work in the optical region of the spectrum, where the atmospheric signatures of oxygen and water vapour were also present. It was to have a mirror four times larger than the Hubble Space Telescope, making it the largest single mirror ever constructed. How something this large could be launched was unclear but, nevertheless, TPF-C won the incoming NASA Administrator's approval. Sean O'Keefe was appointed on 21 December 2001. Less than a year later, he was saying to people like Geoff Marcy that the Terrestrial Planet Finder was to be the coronagraph (TPF-C), not the interferometer design (TPF-I).

This raised planet hunters' hackles because studies had shown that the coronagraph was not as good at seeing Earth analogues as the interferometer. However it could perform a broader range of astronomy beyond planet hunting, making it a more palatable mission for the wider astronomical community. Unsurprisingly, the people who championed the interferometer struck back. They pointed out the technological challenges of building such a large space telescope, its limited application to exoplanets and the fact that if NASA chose the coronagraph, it would have to go it alone because there was no coronography option in Europe. ESA was still steadfastly behind the interferometer.

Whereas the Americans were choosing to start again and dig in a different place every time they came to a rocky layer, in Europe the ethos was to excavate harder and get through that difficult layer. As a result, the more they studied the technological problems with their industrial partners, the more solutions they were finding. Confidence in the design was at its highest level ever. There was a real feeling that Darwin was possible with enough money.

Even so, Fridlund was seeing what was happening in America and was becoming concerned. NASA had been such a good partner to start with; now it was looking unreliable. This was not to say that all Americans were abandoning interferometry. Those teams fought back, pointing out the flaws they saw in the coronagraph design. Distant storm clouds gathered in January 2004, when President Bush gave a speech to define NASA's future role. He chose to focus only on the manned endeavours, calling on the agency to commit to three goals: to finish the construction of the International Space Station, to develop a new crewed vehicle to replace the space shuttle, and to return astronauts to the surface of the Moon by 2020. Then, with those skills, NASA could contemplate sending humans to Mars. He said nothing in the speech about finding Earth's twin planet; that was relegated to just one of around 20 subsequent goals with the directive: 'Conduct advanced telescope searches for Earth-like planets and habitable environments around other stars.'

NASA responded with a document called *The Vision for Space Exploration*, which set out a road map for how it was going to achieve the President's aims. In it, they reaffirmed their interest in building the Terrestrial Planet Finder and pursuing the Origins Program. The document even hinted at missions beyond the TPF, giving them evocative names such as 'Life Finder' and 'Planet Imager'. Importantly, however, the artist's concept of TPF that they used in the document was the interferometer rather than the coronagraph. Subsequently that year, O'Keefe said that NASA would find the money somehow to build both.

Even at the time, no one really believed him and it fell to Charles Beichman to explain the strategy. Although NASA still paid lip-service to the Origins Program, Beichman's title had been changed from Origins Program Scientist to Executive Director, NASA Exoplanet Center. It was a subtle realignment, taking the emphasis off finding our origins and placing it simply on the exoplanets themselves.

When he made presentations about NASA's exoplanet strategy, Beichman referred to himself as TPF Project Scientist. In those lectures, he explained that the coronagraph spacecraft would be in space within a decade. Rather than a giant mirror, it would have a more manageable 4×6 metre mirror, oval in shape. It would be a scouting mission, able to make a relatively quick reconnoitre of the 30–50 nearest Sun-like stars. At the same time, NASA would continue to develop the interferometry mission with ESA, with a planned launch around 2020. This would allow them to survey more than a hundred nearby stars for habitable planets.

It was through international collaboration, he claimed, that NASA would have enough money to do all this. It also meant that NASA might find Earth's twin all by itself if it got lucky with TPF-C and not have to share the limelight. None of this was lost on European scientists.

If further reassurance about NASA's intentions were needed, then the doubtful could look at its preparations for SIM, the Space Interferometry Mission. Hundreds of millions of dollars had been spent on developing the technology needed for SIM. There were labs full of equipment that showed how it would all work. All that was needed was for the mission to be built and launched. Here NASA stood the chance of gaining the upper hand. In Europe, Fridlund and the Darwin team had applied for a technology-testing mission to perform the kind of nulling interferometry that the full Darwin mission would rely upon. This was the miniature telescope mission that would use two free-floating space telescopes, each the size of a tin can, to demonstrate that the technique was viable. But there was not enough money in the ESA kitty and the application was turned down.

For Fridlund, it was a wake-up call. So far, ESA had spent tens of millions of euros on technology development for Darwin. Although it was a lot of money in ESA terms, it was ten times less than would be needed to build a technology demonstrator mission – and the agency had been unwilling to allocate the money. He began to sense that priorities were changing and that perhaps searching for Earth's twin planet was falling from the top of the list.

Instead, he proposed a ground-based nulling instrument that could be used at ESO's VLT telescope in Chile. He called it the Ground-based European Nulling Interferometry Experiment (GENIE) and set about raising the necessary finance. Here too he met resistance, with some believing that it was a backward step for the European Space Agency to be building instruments for ground-based telescopes. So, this proposal ran aground also.

By 2005, Fridlund had a sinking feeling. There was an upcoming call for mission proposals that was going to take place in 2007, but he had a complicated spacecraft design on his hands and no way of testing the hardware or the technique. To safeguard against failure, the agency would demand a high level of elaborate safety systems. These are called redundant systems and are basically a second way to do something, or an exact duplicate, in case the primary system fails. Yet putting more failsafes into the spacecraft would increase its complexity and cost, which would require more safeguards, and Fridlund saw that this path would lead to spiralling costs. Worse, the constant questioning of the interferometer design at JPL meant that NASA was unlikely to formally commit to co-funding the mission.

Indeed, NASA were facing issues of their own. The space shuttle and space station programme were still billions over budget. In 2003, the returning Columbia space shuttle had disintegrated over Texas, plunging NASA into a new crisis. Nothing was going well for the agency. Chief Administrator Sean O'Keefe had resigned after just three years in the job and a new administrator was finding his feet. Mike Griffin was thought to be an excellent choice. Indeed, his first crowd-pleasing decision was to reverse the decision taken by O'Keefe not to refurbish the Hubble Space Telescope. It was the only untarnished jewel left in NASA's crown. The mirror defects of the 1990s were long forgotten and it was seen as an icon. To many, Hubble was the public talisman of NASA. His second popular statement was to tell the nervous scientists who depended on NASA for funding that, despite the human space flight programme facing a budget shortfall of $3.2 billion, 'not one thin dime' would be taken from the science budget to fund human space flight. The statement had the ring of truth because Griffin had started his career as a physicist, so many thought that science ran in his blood.

It was therefore with dismay and a real sense of betrayal that many read NASA's 2006 budget request, which took almost 25 per cent off the science budget. And the cuts did not fall uniformly. Planet searching and the necessary astrobiology to understand the signs of life on alien worlds were eviscerated. SIM was delayed by at least three years while the Terrestrial Planet Finder was 'deferred indefinitely'. This meant that all technology development had to stop because the budget for those missions was reduced to zero. Beichman said he could have kept the programme going with just $10 million out of NASA's $17 *billion*. But there was nothing to be spent on TPF. Instead as much money as possible had to be diverted to the President's wish that NASA rededicate itself to returning astronauts to the Moon.

To make way for this, astrophysics and planetary science missions were removed from the manifest. These included missions to explore Jupiter's moon Europa, to return rock samples from Mars, and to search for gravitational waves in the fabric of the universe. In one of the most inglorious parts of the debacle, Fiona Harrison of the California Institute of Technology and the project scientist of a previously approved NASA X-ray telescope, learnt during the budget press conference that her mission had been cancelled – not deferred, flat-out cancelled. Apparently, no one at NASA had thought it necessary to inform her of the bad news in private beforehand. As the fallout began, many scientists expressed their dismay that the cuts seemed to have been taken on administrative grounds rather than on science priorities. It was reported that

NASA's science advisory council had not been included in the process.[22] As shocking as that was initially, it did mean there was a lifeline.

At the end of the decade there would be the fifth Astronomy and Astrophysics Decadal Survey. Every decade since the 1960s, the National Research Council of the National Academy of Sciences in the United States has supplied a report that surveys the field of outer space research. Based on the responses it received from the various scientists in the field, it would make recommendations about the science priorities for the next decade. It was a possible route to revitalizing the exoplanet missions but, until then, the Terrestrial Planet Finder was in limbo.

NASA wasn't the only team fragmenting over the search for exoplanets. Within astronomical circles, the names Marcy and Butler were fused like neutrons and protons in an atomic nucleus but, to the public at large, that was increasingly not the case. Marcy played well on television and in print. He was unafraid of talking about his emotional journey with depression and the philosophical implications of his science. When talking about exoplanets, he would pluck the kind of inspirational quote out of the air that could elevate a story from a sidebar to a major announcement. When confirming the first planet back in 1995, Marcy said that seeing the world felt like being on Columbus's ship. The implication was clearly about sighting a brave new land for the first time. There are few scientists that can bring this kind of emotional resonance to their interviews – or perhaps only a few scientists who feel comfortable doing this.

Queloz remembers that there was little interest from the

press in their original announcement of 51 Peg b until Marcy confirmed the world and started to work his oratory magic. Then the interest exploded. As a result, Marcy became the go-to man for exoplanet soundbites, even if he had not been directly involved in the research. He appeared on the wildly popular David Letterman chat show in 2001, and his car sports a bumper sticker that reads 'Good planets are hard to find'.

Initially, this disparity in public recognition was fine. In many ways it suited the more introverted Butler because Marcy fed the press while Butler worked the data. But gradually the public recognition turned into academic recognition, honours and prizes. Although some of these did honour them both, there was an increasing tendency for Marcy's to be the only name that appeared. He was elected to the prestigious National Academy of Sciences; Butler was not.

Vogt remembers that a critical moment came in 2005. Marcy had been nominated for, and won, the Shaw Prize. This was a big deal; it is sometimes referred to as the Nobel Prize of the East because it is administered by a foundation in Hong Kong. It rewards outstanding international achievements in astronomy, life sciences and mathematics, and carries a prize of $1 million. Marcy's co-recipient was Michel Mayor. Neither Butler nor Queloz was named.

Marcy, seemingly aware of the sensitivities with Butler, chose not to mention the award. Little did he know that Butler had already been informed privately by a friend who had sat on the judging panel. Marcy flew to Hong Kong to receive the award, issuing a press statement through the Berkeley press office that made constant reference to his team.[23] He stated: 'I feel honored to represent my team for the Shaw award. By discovering over 110 planets so far, our team has started to place our own Earth and its solar system in the context of our grand universe.'

He went on to single out Butler: 'It's a bit embarrassing getting such a big prize. My collaborators really deserve it for all of their innovative and hard work. My long-time buddy, Paul Butler, has been the brains and the engine behind our planet search. I am indebted to him for every single planet we've found.'

Perhaps to underline Marcy's embarrassment, he took just $50,000 of his share of the prize money. Of the rest, he donated $400,000 to the planet search programme at Berkeley and $50,000 to San Francisco State University where he and Butler had begun the research.

Some time after his return, he finally sent an email to his team apologizing for the unfairness of winning the award, and assuring them that he had mentioned them all at the ceremony. Vogt remembers this as being some two months afterwards; he sent a message of congratulations, as did others such as Debra Fischer, an assistant professor of astronomy at San Francisco State University who had also joined the team. But from Butler, who was now a staff scientist at the Carnegie Institution for Science, Washington DC, there was nothing.

The next thing that Vogt recognized was that Butler had stopped coming out to the observing runs at Keck, and he realized that a crack had appeared in the team. Vogt began to feel increasingly uncomfortable, and in the autumn of 2007 he made the decision to leave. He told the rest of the team that he was just going to withdraw and let them get on with it. His message elicited an unexpected response. A day later, Butler broke his silence. He too said that he was leaving the team, but that he wanted to work with Vogt. This was no longer a question of one person leaving a team; now it was a full-blown rift. To those involved and to other exoplanet scientists, this event is still referred to as 'the divorce'.

As with any divorce, the assets must be split. In the case of this team, the most obvious assets were the unanalysed data. It was inconceivable that Butler or Vogt could simply walk away. Instead, the data that they had helped take had to be divided. They discussed how to do this and decided that they would draw lots for the thousands of observations they had on record. So the stars in their programme were divided in this way. Each side agreed not to observe the other's targets for a full year, allowing them to analyse the data, take any other measurements they needed and then publish the results. After that, the teams would be in competition.

There was also a bigger prize. Thanks to the assistance of a political lobbyist, in 2004 Congress had signed into law that the team should receive $8 million from the Department of Defense's budget to build a planet-finding telescope. The Automated Planet Finder was to be a 2.4-metre telescope, sited at the Lick Observatory, and it was only partially complete. Vogt set about completing the project and now split the time on the telescope evenly between him and Butler, and Marcy.

At the time that the breakup was happening, there was acrimony as Vogt remembers it. Although Marcy does not downplay this aspect he now says that a split was inevitable at some point because students never work in their supervisor's shadow for too long. They need to spread their wings and develop programmes of their own, with students of their own. He has been quoted as saying that the reason he prevaricated in telling his team about the Shaw Prize is because he feared that the inequity of it would cause the divorce it did.[24]

By late 2007, it became obvious that other political lobbyists were at work around Capitol Hill. Congress directed NASA to spend a full $60 million on the SIM mission, almost three times more than the agency had planned. An angry Mike Griffin spoke at a meeting of the American Astronomical Society in Austin, Texas, early in January 2008. In a dressing-down reported by *Nature*,[25] he admonished the assembled astronomers, 'I hope this is what you want because it appears likely to be what you will get.'

He went on to say, 'Congress does not dream up such directions on its own. Clearly, external advocacy for SIM has been successful.' Griffin and other managers at NASA warned against the danger of lobbyists gaining support for individual missions. Flagship projects, such as the James Webb Space Telescope, could be delayed, they said, or even cancelled. Their worry was that the dwindling science budget could probably not sustain two $1 billion missions. Astronomers were likely to face a straight choice between SIM (see Chapter Five) or JWST (also see Chapter Five, when the mission was known as NGST): a specific exoplanet mission for a few astronomers or a general space observatory for all. Clearly, finding Earth's twin was no longer a priority for the agency.

In Europe, things were also turbulent. The loss of NASA as a partner and continued concerns over the cost of Darwin meant that ESA quietly put the project on ice. It was not a cancelled mission like Eddington because it had never been commissioned in the first place. It had only been studied. While the agency said that it retained the science goals of

Darwin, the titular mission was impractical. For Fridlund, it was a depressing time. He was seriously thinking of leaving ESA. He had worked there since his Ph.D. yet none of the missions he had studied had been chosen to fly. His bosses realized the depths of his disaffection, however, and also that his combination of talents and experience was exactly what they needed for CoRoT. They asked him to become the mission's project scientist.

It changed everything for Fridlund. In his soul-searching about the shortcomings of the Darwin mission, he had zeroed in not on the technology – which he maintains was realistic and would have cost about $1.5 billion for ESA to build on its own – but on the assumptions that had gone into the investigation. The science team was simply taking it on faith that every nearby Sun-like star had an Earth-like planet because they had no other knowledge with which to fill that gap.

So Fridlund had concluded that, before some Darwin-like mission was resurrected, astronomers had to first know what the frequency of Earth-like planets was around nearby stars. The missions that could do that most easily were CoRoT and Kepler. So, being placed in charge of ESA's involvement in this largely French mission suited him perfectly.

CoRoT left Earth courtesy of a Russian Soyuz rocket on 27 December 2006. Its journey into orbit was flawless, and the space telescope was collecting science data by 2 February 2007. Within a few months of operation, the spacecraft bagged its first planet. CoRoT-Exo-1b, as it was called at the time, was a classic hot Jupiter, with a diameter 1.78 times that of our own Jupiter and an orbital period of just 1.5 days. It was located around a star 1,500 light years away in the direction of the constellation Monoceros. Follow-up radial-velocity measurements from the ground had confirmed the planet's existence by seeing the star wobbling, and this had given the planet's mass as 1.3 Jupiter masses.

As was now customary, the astronomers calculated the average density by dividing the planet's mass by its volume, and the result was an eye-opener. CoRoT 1b, as the name was soon shortened to, had a much lower density than the gas giants in our solar system. It joined the ranks of the first detected transiting planets, OGLE-TR-10 b and HD 209458 b, in this regard, and astronomers knew that they had some explaining to do. Something was puffing up these planets' atmospheres and lowering their density to around 0.35 grams per cubic centimetre, almost one-third the density of water.

Researchers first thought that, being so close to their central star, these worlds will be heated so much that the hot atmosphere will naturally expand, but the sums didn't add up. The worlds were too large to be explained solely in this way. The current thinking is that it is a magnetic interaction called ohmic heating that is taking place. Both the central star and the hot Jupiter will be powerfully magnetic, and the interaction of these magnetic fields will create further heat, inflating the atmospheres to their observed bloated, low-density states.

As intriguing as the science was, there was even better news coming from the engineers. CoRoT was working spectacularly well: much better even than the design specification. When all of the sources of error were taken into account, it was going to be possible to see planets down to the size of the Earth, three times smaller than the mission was expected to discover. Fridlund and the science team realized that this meant it was possible that they could scoop NASA's Kepler by being the first to discover Earth's twin.

By October 2007, the spacecraft was operating smoothly and routinely. On 15 October, it began observing a Sun-like star about 500 light years away in the constellation Monoceros. The data was automatically returned to Earth, where computers analysed the signals. With 40 days of data under their robotic belts, the computers alerted their human masters that there could be something worth investigating around the star, which the team called CoRoT-7. The computers had identified what looked like 153 transits, each one lasting around 1.5 hours. The orbital period was really short, just 20 hours, but the most important thing was the depth of the transit. The star was dimming just 0.03 per cent, thus indicating a small planet indeed. The time taken for the star's light to drop into the full transit showed that the planet's diameter was just 1.58 times the size of the Earth – much smaller even than the super-Earths. If confirmed, CoRoT-7b would clearly be the first planet that could legitimately called Earth-sized.

The CoRoT team contacted Didier Queloz, who was now the de facto head of European planet-hunting efforts because his supervisor and colleague Michel Mayor had officially retired in 2007 not long after sharing the Shaw Prize with Marcy. Queloz immediately took up the challenge and got to work with the HARPS instrument. Collecting the data was easy – Queloz had a running programme of observations – but the analysis proved more difficult. The star itself was much more active than the Sun, with large dimmings and flares. Working with the data as best they could, Queloz and his team extracted a probable mass for planet CoRoT-7b of about 4.8 Earth masses.[26] Combining this with the volume, as given by the diameter, meant that the average density of the planet was similar to Earth's. It was indeed the first assuredly rocky planet to be found by the exoplanet hunters. It was announced in February 2009 at a special conference devoted to the findings of the CoRoT space telescope.

Inevitably the discovery of a rocky planet generated great interest in the media and with other scientists. It was clear from the very beginning that the planet was going to be another hellish environment. It was so close to its central star that the rocky surface would be heated to a temperature somewhere between 1,800 to 2,600 degrees Celsius. As rocks melt at temperatures in excess of 1,200 degrees Celsius, this world could well be covered by oceans of lava rather than water. It was also highly likely that tidal forces from the star had locked CoRoT-7b into presenting only one face to the star. In this case, the near side could be a giant lava ocean but the far side, bathed in perpetual night, played host to a frozen wasteland of ice that had once been the planet's atmosphere. It was an amazing world to contemplate, but it was not Earth's twin, and now the Europeans had competition because on the other side of the Atlantic, the Kepler space telescope (see Chapter Five) was on the pad at Cape Canaveral, Florida, ready for launch.

During all of the budget shenanigans, Kepler's Bill Borucki had kept his head down. Working at NASA's Ames Research Center, Borucki saw the worst effects of the budget cuts because much of NASA's astrobiology work was based there. With funding all but drying up for that endeavour, around 100 contractors lost their jobs. Many of the victims were young researchers trying to kickstart their careers. Borucki's Kepler mission was safe because it was far advanced and he was keeping it steadfastly on time and budget. Now, the $600 million spacecraft was sitting atop a Delta IV rocket, ready for its accent to orbit. At a press conference before launch, Borucki emphasized that the goal was finding Earth's twin. He told reporters, 'We certainly won't find E.T., but we might find E.T.'s home by looking at all of these stars.'

What he did not mention was that, in the final stages of the build, the team had encountered a technical issue that could jeopardize the entire mission. It had come to light two years earlier when the spacecraft's manufacturers, Bell Aerospace in Boulder, Colorado, developed misgivings about the reaction wheels.

Reaction wheels are hand-sized spinning discs that keep the spacecraft pointing steadily in the right direction. The precision of the pointing was essential to get a steady signal from the stars under observation and this required three reaction wheels. Without a full set, the spacecraft would wobble and this 'noise' would utterly overwhelm the planet signals.

The reaction wheels work because of Newton's third law of motion, which states that for every action there is an equal but opposite reaction. The spacecraft moves in the opposite direction to the spin of the wheels but because the spacecraft is much larger and more massive, it spins gradually rather than at the wheels' furious rate of a few thousand revolutions per minute. The onboard computers sense any stray motion and spin the reaction wheels in the appropriate way to counteract the movement. From experience, spacecraft engineers know that moving parts in space often wear out quickly. So they had installed four reaction wheels on Kepler, effectively providing a spare. But these particular reaction wheels seemed to be wearing out faster than expected.

Bell Aerospace's concerns had been triggered by the failure of a similar component on board NASA's Thermosphere, Ionosphere, Mesosphere Energetics and Dynamics (TIMED) satellite, which had been launched in 2001. Further investigation[27] showed that almost identical reaction wheels had failed on Japan's Hayabusa asteroid mission in 2005. With launch fast approaching, Bell Aerospace and NASA had to do something, and fast. The obvious solution was to replace it with a new system, or add in more reaction wheels for additional backup, but with the spacecraft essentially complete either option would have cost a lot of money and delayed the launch, which would itself have cost more money. This was not a good option in the new budget-restricted times. Also, the spinning reaction wheels vibrated the spacecraft, imprinting noise on the data. This was well understood for the chosen reaction wheels, and sat below the threshold that would impact the telescope's accuracy. A different set of wheels would rumble in a different way and almost certainly degrade the performance of the telescope.

It was the proverbial rock and a hard place. So Bell and NASA compromised on refurbishing the original system and refitting it as if it were new. Having done this, both were confident that the wheels would last for the nominal three-year mission. Nevertheless, all knew it was likely to be the spacecraft's Achilles heel. Undeterred, the science team had identified a star field near to the plane of our galaxy that crossed several constellations: Cygnus, Lyra and Draco. It represented about 1/400th of the whole sky and let the spacecraft monitor 150,000 stars simultaneously, 90,000 of which were Sun-like stars. The science team had agreed that no announcements would be made until they had seen a planet make at least three transits of its parent star. This was only sensible. The first transit alerted them to the planet's existence. The second allowed them to measure the length of its year. This is turn allowed them to make a prediction about when they would see the third transit. If they saw the star dim on schedule, then they agreed that an announcement of a discovery would be made. So to find Earth's twin was going to take at least 2–3 years, since that would indicate a planet in an Earth-sized, 12-month orbit.

Following launch on 7 March 2009, Kepler's first exoplanet discoveries were announced to the press at a meeting of the American Astronomical Society in Washington DC on 4 January 2010. There were 125 possible planetary signals in the data so far, explained a clearly delighted Borucki. Of these, five had been confirmed with radial-velocity measurements from Earth. Four of them were puffed-up hot Jupiters. One of these, called Kepler 7b, was the most insubstantial planet yet, with a density similar to styrofoam.[28] The fifth planet was a hot Neptune. All the planets circled stars that were somewhat hotter than the Sun and so were roasted to temperatures of just below 2,000 degrees Celsius.

Borucki also reported that the telescope was working excellently; everything was on course to find Earth's twin. The good news was particularly timely because as Kepler was launching and the scientists were collecting the first data, astronomers from across the United States were submitting their 'white papers' to the National Academy of Sciences as part of the decadal review of priorities in astronomy. It was no secret that a number of those papers were about the Terrestrial Planet Finder. This was the opportunity to get it back on the agenda.

No one, it seemed, anticipated the disaster that was about to happen.

It came in the form of the decadal review's 290-page report entitled *New Worlds, New Horizons in Astronomy and Astrophysics*. Ten years earlier, the previous report had mentioned TPF on 22 pages. Only the James Webb Space Telescope, referred to back then as the Next Generation Space Telescope, was mentioned more times. Yet in the 2010 report, TPF was mentioned just once, in a summary of what NASA had been doing based upon the previous report. There was no recommendation to recommence the mission, either as an interferometer or a coronagraph.

Worse was the treatment for SIM, which was now referred to as SIMLite to indicate a slightly scaled-down version. It was explicitly mentioned just once, in a footnote rather than the main text, which effectively cancelled the mission.

The footnote reads: 'In considering possible exoplanet missions for the next decade, the committee gave serious consideration to SIMLite but decided against recommending it. SIMLite is technically mature and would provide an important new capability (interferometry). Through precision astrometry it could characterize the architectures of 50 or so nearby planetary systems, provide targets for future imaging missions, and carry out other interesting astrophysics measurements. However, the committee considered that its large cost (appraised by the CATE process at $1.9 billion from FY2010 onward) and long time to launch (estimated at 8.5 years from October 2009) make it uncompetitive in the rapidly changing field of exoplanet science.'

It was pretty damning stuff, directly contradicting the

project scientist, who said he thought the mission could be built for a billion dollars. But the most embarrassing thing was that the report made reference to a fundamental split within the exoplanet community over the need for SIMLite. As originally envisioned, the mission was designed to test the concept of interferometry in space by targeting 50 or so nearby stars. The precision at which SIMLite could work was expected to reveal multiple planets around these stars, which the researchers called planetary architectures. In this way SIMLite would find the targets for the Terrestrial Planet Finder. So to those who still favoured the interferometer design it was an essential first step. However, the scientists and engineers who wanted the coronagraph version of TPF saw it completely differently and sought to distance themselves from SIMLite. They claimed to be able to find and analyse planets in one fell swoop. If SIMLite was cancelled, it would not affect them at all. This civil war among the researchers had come through loud and clear to the committee of the decadal review.

But there was also an embarrassment for the report writers themselves on the very next page. In summarizing the recommendation for how to implement the 'New Worlds Science Plan' as they called it, one of their bullet points said 'Follow up nearby systems discovered by Kepler.' But the reality was that Kepler was never going to find nearby systems; its field of view was fixed far away towards dense star clouds so that it could simultaneously monitor many stars – in the same way that you can see every person in a crowd if they are standing on a distant hill but not if they are surrounding you. Every planet discovered by Kepler was going to be at least hundreds, more likely a thousand or more, light years away.

The exoplanet community were chagrined. From having NASA at its feet, now the mighty space agency had all but turned its back on them. In May 2011, the frustration spilled over at a one-day conference organized by Canadian-American astronomer Sara Seager at the Massachusetts Institute of Technology. The title of the conference was 'The Next 40 Years of Exoplanets'. The website[29] showed children in silhouette against a twilight sky pointing upwards, with the caption, 'That star has a planet like Earth. We want to show our children, grandchildren, nieces, and nephews a dark sky, point to a star visible to the naked eye, and tell them, "that star has a planet like Earth." We will make this possible within the next 40 years.' With typical academic rigour the words 'like Earth' had been asterisked. The explanation was given as 'Earth-like means a rocky planet in an Earth-like orbit around a sun-like star that has strong evidence for surface liquid water.' Earth's twin. It was clearly designed to be a rallying cry for exoplanet hunters to recommit to the search. It turned into something of a blame game.

People were still filing back into the room after coffee break on the first day as Seager introduced Marcy with the classic 'this person needs no introduction' introduction, and Marcy was remarkably frank from the word go. Speaking with his hands planted in his pockets he first said that he was 'ecstatic' about the Kepler mission and the results but he was also 'unhappy about the last ten years and unhappy at the next ten years'. He promised to use his talk to describe his 'anger'.

First, he explained that all flourishing fields of astrophysics relied on spectroscopy to provide the kind of data that could be used to extract precise physical and chemical conditions of the objects under scrutiny, yet this was the one thing that now seemed denied to the exoplanet hunters. He reminded the audience that spectra would have made it possible to study the constitution of planets and by extension whether they could be habitable. This should have made the case for TPF extraordinarily compelling, he said.

At the beginning of the 21st century, he explained, the study of exoplanets was somewhat similar to the study of stars at the beginning of the 20th century, when it had been effectively a game of discovering stars and counting them. This was essentially what the exoplanet hunters were doing now. They could speculate a little about the physical characteristics of the planets but, to make any real progress, they needed to study them spectroscopically.

He emphasized that the nearby stars were the future of exoplanet research. They are easier to see, they can be imaged, allowing spectroscopy, and they are the ones that will be eventually reachable with 'next-generation' space probes.

He turned next to the 2010 Decadal Survey. Despite the fact its title included the phrase 'new worlds', how come it did not recommend TPF, he asked rhetorically. He thought that it happened because the exoplanet community did not unite to voice a single coherent message about the need for such a mission. It was unfathomable to Marcy how the exoplanet community could have let this opportunity slip through their fingers.

'We went from interferometer to coronagraph to both to nothing,' he said in a voice of incredulity.

The reason for this abrupt reversal of fortune, he proposed, was that there was no strong leader who emerged to keep the in-fighting under wraps. He said that the leadership could have come from NASA headquarters in Washington DC, or

JPL in Pasadena. He partly blamed himself for not taking a stronger role. Somehow, instead of feeling like part of the same project, with the final spacecraft design a matter for internal discussion only, the different advocates fought for supremacy of their own pet design, as if the spacecraft was the 'thing' not the resulting science. As if that wasn't bad enough, the various teams' method of doing that was not to highlight the potential of their own missions but to point out the weaknesses in their rivals. It was the scientific equivalent of negative political campaigning and this very public squabbling had damaged the credibility of the entire field. Without TPF in the decadal survey's recommendations, there could be no progress.

Marcy turned next to NASA, reminding the audience that the space agency had allocated zero budget for technological development, zero budget for selecting the appropriate mission design, and zero budget for investigating the basic science drivers for TPF. How could NASA do that after their leading efforts of the last decade? He said simply that NASA was just 'jerking them around' by vacillating between coronagraphs and interferometer versions of TPF.

Way back in 1999 he had sat on the TPF technical team and had seen the interferometer design begin to emerge. It was ambitious for sure but there was more at stake than just exoplanet science. He nailed his colours to the mast by saying that the only plausible future for astrophysics, whether exoplanet or any other branch of the discipline, was to embrace space-borne interferometers like SIMLite and TPF/Darwin. Interferometry was the easiest, most obvious route to seeing the universe in the finest possible detail.

Astronomers who study the universe at radio wavelengths

have known this for half a century and now routinely build and use whole fields of separate dishes, all feeding into each other. It was time for the optical and infrared community to catch up. In one particularly angry moment, he said that the proponents of TPF-C had really 'pissed him off' because their attempted coup had derailed this effort almost completely. There was no laboratory equipment that could back up their claims to do better than the interferometer, he said, yet there was a complete lack of humility to honestly talk about the uncertainties in the technical design. And so the field was stuck until at least the next decadal review in 2020.

At the end of the session, Wesley Traub from NASA stood to respond. He had joined NASA's Jet Propulsion Laboratory in 2005 to become the project scientist for TPF-C. About a year later, he found that the funding was drying up and two more years then went by before he was given the more generic title of NASA's Chief Scientist for their Exoplanet Exploration Program. It would have been easy for him to feel aggrieved at Marcy's criticism of weak leadership yet, when he spoke, he thanked Marcy for getting down to the brass tacks of the matter.

He said that he had thought about the problems a lot because he was now in a position where 'I'm having arrows fired at me, almost constantly, or at my organization – and with good reason too, I think.'

He admitted to a sense of pride at helping to develop TPF-C, because it seemed like an easier approach, just a single telescope rather than many. He did not have anything against interferometers, he said, but since working at JPL he had found that there was a prejudice against them. It was more to do with perception and sociology than science, and he admitted that it was discouraging to find such unscientific prejudice at a scientific establishment.

Interferometers are perceived as difficult-to-build, niche instruments. Missions these days, said Traub, had to offer 'full employment' for astronomers: in other words, a general observatory that could be used across the broad sweep of astrophysics. About the SIM cancellation, Traub was blunt. 'It was the most humiliating, embarrassing thing I've ever seen in my life, a footnote in fine print dismissing $600 million worth of work that had been advocated by two previous decadal surveys.'

To set this money into context, Marcy had earlier said that six entire Keck telescopes could have been built for the same money NASA had poured into SIM before cancelling it. A seventh, with cash left over, could have been made with the $120 million NASA had spent on Keck interferometry as a precursor to TPF, which was now also junked.

Traub's take was clear that the decadal review was a self-serving document that sought to retain full employment for people to continue doing what they were already doing. In other words, rather than push the boundaries of science, it was a steady-as-she-goes situation. His final point was to mildly admonish the community by saying that they should stop talking about TPF as a complicated, $5 billion mission. According to Traub, JPL had laboratories full of experiments that showed how the mission could be achieved. 'It could be done today, all we have to do is start work on it.'

This shocked audience member David Carbonneau, who had been one of the astronomers who had discovered the first transiting planet in 1999. He too addressed the audience. He said that he supported the decadal survey despite the frustration it caused because it was the only possible outcome given the white papers that had been submitted by personnel from JPL.

He had served on the panel for planetary systems and star formation and so had read all the white papers. Part of the frustration, he said, is that JPL – one NASA centre – had submitted two or three white papers with radically different proposals for doing the same mission. He said that it was like an animal trying to cut its leg off from a trap. The TPF-C people were saying we don't need SIM because they could tell that the mission was in danger. So if you made your own mission reliant on it, then you were automatically cancelled if SIM went by the board.

Then there were the papers from TPF-I supporters saying you can't do this with a coronagraph. Carbonneau asked: How can one NASA centre have so many different points of view, and spend so much of their white papers talking about how other ideas will not work?

Traub did not rise to the bait, and Charles Beichman, who had been TPF's project scientist, was not at the meeting. Later, in explaining what happened at that time, Beichman said that the plain truth of the matter was that, even among the exoplanet community, not everyone was interested in searching for life on Earth-like worlds. Many of those astronomers simply wanted to study the whole diversity of exoplanets that were out there in order to understand more about the way planets form. They were interested in the physics of planet formation rather than any possible biology that was going on. In a strict sense, it was probably the more scientific attitude to have but it meant that they didn't fully get behind TPF or Darwin to help push the missions forward.

Seager concluded that part of the workshop by commenting, 'There is a saying about "divide and conquer": we divided and got conquered.'

With TPF off the table, all eyes turned to Kepler, CoRoT and the various ground-based surveys to find Earth's twin. Finding the first habitable planet was a race into the history books, and the headline wars were about to begin.

EIGHT

Headline wars

The first planet that astronomers felt bold enough to call 'Earth-like' was Gliese 581 c (often shortened to Gl 581 c). It was discovered by a Swiss team, including Mayor, using the radial-velocity 'wobble' method, and was announced in 2007. At first glance, the title of the paper was interesting but not exactly exciting. Posted on a Sunday to the ArXiv website, a repository for scientific papers that are undergoing the sometimes long-winded process of review for publication, the paper was called 'The HARPS search for southern extra-solar planets XI. Super-Earths (5 & 8 M_Earth) in a 3-planet system'.[30] Those in the know would have immediately spotted that the 5 Earth mass planet was the smallest world so far found. The six-line abstract, however, considerably upped the ante by revealing that the planet sat at the 'warm' edge of the habitable zone of the star. This, and the planet's mass, emboldened them to claim, 'It is thus the known exoplanet which most resembles our own Earth.'

Headlines followed. Space.com ran with 'Major Discovery: New Planet Could Harbor Water and Life'.[31] It quoted lead author, Stephane Udry, as saying that when he calculated the planet's likely temperature and found that it was probably between 0 and 40 degrees Celsius, and thus capable of having running water at its surface, he knew that he was going to be fielding a lot of enquiries from the press. 'You right away think about the journalists who will like it very much,' he told the website.

Perhaps that was why he placed such a provocative sentence in the abstract. Clearly his collaborators backed him up. One of them, Xavier Delfosse of Grenoble University in France, told Space.com, 'On the treasure map of the universe, one would be tempted to mark this planet with an X.' It was something of an overstatement, of course, based largely on speculation, but it did mark a significant step along the path to finding Earth's twin.

Dig into the details though, and the planet did not sound very Earth-like at all. First, its mass was 5 times larger than the Earth, meaning that its diameter would probably be about 1.5 times larger. Its orbital period was 12.9 days, placing it just 0.0721 of an astronomical unit from its parent star. Although this was much closer than the inner boundary of the Sun's habitable zone, the star was not like the Sun. It was a fainter red dwarf star, less than one-third the mass of the Sun. It was just over a quarter of the Sun's diameter and produced a feeble light that carried a total energy of about one-tenth the Sun's output. This is why the planet had to be so close to the star, in order to compensate for the lack of intensity. The price for this proximity was that the planet would be tidally locked, meaning that one hemisphere would be perpetually heated while the other faced the deep freeze of outer space. The astronomers speculated that any true habitable region on the planet would probably be confined to the margins between the perpetual day-lit side and the dark side of the planet. Even this might be challenging, though, because the vast temperature difference between the planet's two hemispheres was likely to generate ferocious winds.

The biggest fly in the ointment for the team was that the astronomers had not detected the planet transiting. This robbed them of the planet's diameter, from which they could have calculated the average density, to see if it was consistent with a rocky planet. Without this, they could not say for sure whether the planet was a rocky super-Earth, or a gaseous mini-Neptune. As the astronomers thought more about this, however, they decided that if Gl 581 c were a mini-Neptune, it opened an even more exotic possibility. It could be an ocean world.

In our own solar system, Neptune is thought to contain a rich mix of icy material under its thick atmosphere. If it were to migrate in towards the Sun, much of the ice in the planet's interior would melt because of the higher temperature, covering the planet in a global ocean hundreds of kilometres deep. Contrast this to Earth's oceans, which have an average depth of about 3.5 kilometres, reaching down to just 11 kilometres in the deepest parts.

Given the wind-scoured twilight realms postulated for a rocky planet, it may be that the chances of finding life on Gl 581 c are much better on a water world. However, the planet would be tidally locked and so the maelstrom of the atmosphere would be the same. Underneath the almighty waves that these hurricane-force winds would raise, any life forms would be protected. Also, the buoyancy of the water would offer respite from the stronger gravity generated by the mighty world.

The fact that it was the most Earth-like planet yet discovered should really have served to underline how alien the other 230 worlds discovered since 1995 were. But of course, in the public mind, saying Earth-like conjured images of our own world replete with blue skies, crystal-clear oceans and life.

Although it was the 5-Earth-mass inner world that was catching the headlines, climatologists from the Potsdam Institute in Germany were starting to be drawn towards the other planet that had been presented in the paper. The 8 Earth-mass Gl 581 d orbits its star at 0.25 AU, taking almost 67 days to complete its year.

W. Von Bloh and colleagues looked at the insulating effects that such a large planet's atmosphere would have on the planet's surface conditions and, according to their analysis,[32] Gl 581 d was much more likely to be habitable. But, by then, the press had moved on.

The HARPS team led by Francesco Pepe reported super-Earths around HD 20794, HD 85512 and HD 192310.[33] These included one that sat at the inner edge of its star's habitable zone and was just three and a half times more massive than Earth. They did not make a huge announcement at the time but, surely, the detection of Earth's twin was just around the corner.

In 2010, the world's attention was turned back to the Gl 581 planetary system. Since deciding to go it alone, Butler and Vogt had been adjusting to planet hunting without Marcy. And vice versa: Marcy was busy building a new team. One of the stars that Butler and Vogt had inherited from dividing up the data was Gliese 581. They had 240 nights of observations, much of it collected by themselves over 11 years at the Keck telescope. They augmented this with publicly available HARPS data that the Swiss had used to discover Gl 581 c and d, and when they did the number-crunching, they saw something. It was a faint signal to be sure, but the more they looked, the more convinced they became. By September 2010, they were ready with an announcement of their own.

Their new-found world, Gliese 581 g, was between 3 and 4 Earth masses, making it one of the smallest known exoplanets at the time. Crucially, its orbital period was 37 days, placing it smack between Gl 581 c and d, and that put it right in the middle of the star's habitable zone. The National Science Foundation, an independent federal agency created by the United States Congress in 1950 that funds about a quarter of all university and college research in America, organized an hour-long video press conference for Butler and Vogt to discuss why this find was special.[34]

Both were in bullish mood. Straight off the bat, Vogt said that Gl 581 g was the first exoplanet of the roughly 500 that had now been discovered 'that really has the right conditions for liquid water to exist on its surface'. He went on to say that the two previous worlds that had been discussed as habitable by the Geneva-based researchers had turned out probably not to be so habitable after all. So the message was clear: this was the first unequivocally habitable world.

Then Butler talked and rather pointedly, for those reading between the lines, said that his whole career was 'based upon spectrometers that Steve Vogt had designed and built'. There was no reference to Marcy at all, even when Butler was talking about the origins of the project. When referring to their discovery, Butler called it the first Goldilocks planet, in other words: not too hot and not too cold.

They turned next to the conditions on this world. As was becoming common for such planets close to their parent star, it would be tidally locked. To drive this image home, Vogt spoke of how the planet would not have days or nights. Instead, the star would appear fixed in place in the sky (or not at all if you were on the far side).

Along the terminator, the astronomical word for the line between day and night, the star would be sitting on the horizon forming an eternal sunrise or sunset, according to whether you were an optimist or pessimist, joked Vogt. This meant that the planet would exhibit different 'eco-longitudes', he said. If you were a creature that evolves to like a hot climate, you could move into the areas where the Sun is higher in the sky. If you were the alien equivalent of a polar bear, you could move further round to the cooler side. This was all hypothetical of course, but there was little to no caveating in the presentation.

When discussion turned to funding, they had some similarly interesting statistics. Vogt estimated that observing with the Keck telescopes cost about a dollar a second. With each individual observation of a star lasting for 600 seconds, that meant the price tag for finding a planet was somewhere around $100,000, much of it coming from private investment, and the rest coming from NASA and the NSF. Butler pointed out that was a cheap method of finding planets compared with space missions.

When pressed by journalists about whether habitability meant inhabited, Butler admitted that such talk was speculative but he also pointed out that everywhere on Earth where there is water, there is also life. Because of this, he said, the question should actually be how can you rule out that life *doesn't* exist on Gl 581 g. Vogt backed up this point saying that life began on Earth as soon as it could, making him think that the origin of life was an aggressive, possibly easy process once conditions were right. 'It's pretty hard to imagine that there would not be water there,' he said.

When asked to put a figure on the chances of life, Butler demurred.

'I'm not a biologist, nor do I want to play one on TV,' said Vogt before going on to give his personal opinion. 'The chances of life on this planet are 100%.'

And with that out of his mouth, he guaranteed the story would make headlines across the world.

Across the Atlantic, the eyebrows of those on the Swiss team began to rise. How could they have missed this planet? They had the most sensitive spectrometer in the world. Although they were using it on a smaller telescope than the Keck, Vogt himself had said in the NSF press conference that the Swiss data had been crucial in finding this planet. It was probably meant as a compliment but it sounded a bit like a slap in the face. Immediately, Queloz and colleagues began to reanalyse their data. Early in October, they made an announcement: they could find no evidence to back up the American claim regarding this planet. There was simply no trace of it in their data, even now they knew what they were supposed to be looking for. Clearly, the implication was that Butler and Vogt had mistaken random noise for a planet.

Vogt stood by the discovery, even after other research teams reanalysed the available data and found nothing. One even suggested that planet g and planet d were mistakes because both teams had been fooled by stellar activity. Subsequently, the Swiss team released an updated dataset to Vogt. While the Swiss again found nothing, Vogt did. How could the two teams look at the same data yet come to such different conclusions? As they tried to unpick the mess, it turned out that Vogt had imposed an assumption that all planets in the system were following circular orbits. He thought this entirely reasonable, whereas the Swiss preferred to let the data speak for itself. Vogt hit back saying that he could only reproduce the Swiss null result if he purposely excluded some data points. Although discarding clearly erroneous signals is standard scientific practice, Vogt claimed that the Swiss team had not admitted to leaving out the data, which turned out to be the very signals that revealed the planet.

There could be no doubt that, by this time, temperatures were rising. It did not help that Butler and Vogt were also embroiled in another disagreement with the Swiss team, this time about the potential habitable planet Gliese 667C c. The planet had emerged from a reanalysis of publicly available HARPS data made by Spanish astronomer Guillem Anglada-Escudé. Working at Queen Mary University, London, he had developed new software to look for planetary signals in the discarded data of the major research groups. He had begun collaborating with Butler and Vogt after gaining a postdoctoral position at the Carnegie Institution of Washington, where Butler was a professor.

The Gliese 667 system itself was an exotic locale. It was a triple star, composed of three red dwarf stars. The two larger components, A and B, orbited each other. Way out at a distance of 230 astronomical units, Gliese 667C orbited both of them. It was around this third star that Anglada-Escudé found a planetary signal. With an orbital period of 28 days, it was right in the star's supposedly habitable zone, and with a mass around 4 Earth masses, Gl 667C c was every bit as Earth-like as Gl 581 g. With the latter planet now in dispute, this new one might just be the one to go down in history as the first detected habitable world.

When calculating the amount of energy the planet receives there is a difference between the overall quantity of energy and the quality of the light. Being a red dwarf star with a temperature of about 3,600 degrees Celsius means that the amount of illumination that the planet receives is only about 20 per cent of that received on Earth, and it will be a deep red compared with sunlight, which comes from our star's 6,000-degree-Celsius surface. To our eyes, the world would be a twilight realm of rose-coloured hues. The majority of the 'light' arriving at the planet would be in the form of invisible (to us!) infrared rays. Just as our eyes have evolved to be more sensitive to the peak emission of the Sun, in the yellow part of the visible spectrum, any life on Gliese 667C c would probably evolve eyes more akin to thermal security cameras, detecting the world around it by the heat it radiates.

This time no one queried the existence of the planet; instead they argued over who had discovered it. Soon after Anglada-Escudé made his discovery, a paper appeared on the Internet from the HARPS team, led by new member Xavier Bonfils, University of Grenoble, France. It was a summary of HARPS observations between 2003 and 2009 and it listed Gl 667C c.[35] Vogt saw the paper and shot off an email to Anglada-Escudé saying, 'We've been scooped.'

Early in 2011, NASA's Kepler mission started to show what it was really capable of. The mission had been designed with the express intention of finding Earth analogues, and the mission presented its first rocky planet discovery in January at the American Astronomical Society's winter meeting in

Seattle. The planet was in orbit around a star that the team had called Kepler 10. It was the smallest exoplanet yet discovered, with a radius of 1.4 times the Earth. In truth, it was really just an incremental step along from what had already been discovered but the scientists played it up. The NASA press release contained quotes from the key players.[36]

Natalie Batalha, Kepler's deputy science team lead at NASA's Ames Research Center, California, and the primary author on the Kepler 10b discovery paper said, 'All of Kepler's best capabilities have converged to yield the first solid evidence of a rocky planet orbiting a star other than our sun.'

Douglas Hudgins, Kepler Program Scientist at NASA headquarters, Washington DC, said, 'The discovery of Kepler-10b, a bona fide rocky world, is a significant milestone in the search for planets similar to our own.'

Kepler team member Dimitar Sasselov, of the Harvard-Smithsonian Center for Astrophysics, Cambridge, Massachusetts, said, 'This planet is unequivocally rocky, with a surface you could stand on.'

Again and again the language was chosen carefully to make the press think that this was the first rocky planet that had been found: 'First solid evidence', 'bona fide rocky world', 'unequivocally rocky'. Perhaps the reason that the researchers felt justified in this approach is because most of the previously announced 'rocky' worlds had not been detected transiting. From the super-Earth Gliese 876 d to the almost Earth-sized Gliese 667C c, previous astronomers had only ever detected radial-velocity measurements, giving a likely mass. Without a transit to give the size of the planet, the mass could not be turned into a density, and so some sliver of doubt had to remain about whether the planet really was a compact rocky world or some rarefied, gaseous body. This time round, the Kepler scientists had not just the size, but also the mass thanks to Geoff Marcy.

As part of the Kepler science team, it had fallen to him to confirm the planet's existence by performing radial-velocity measurements using the Keck telescopes in Hawaii. These showed that the planet was 4.6 Earth masses. Combined with its size, it was clear that this was not a gaseous object. It was solid. With typical aplomb, Marcy told the press that Kepler 10b was a 'missing link' in the search between gas giant planets and Earth-like worlds and that the exoplanet, 'will be marked as among the most profound scientific discoveries in human history'.

While this game of intellectual cat and mouse could almost be justified, NASA and the other researchers involved seemed to have conveniently forgotten about CoRoT-7b from 2009. It had been discovered by its transit signals, and then confirmed by radial-velocity measurement. Although the star was particularly active and this made measuring an accurate mass difficult, there was no real doubt that it was a solid rather than a gaseous world. So to claim the Kepler 10b was the first was simply not accurate. Nevertheless, the press ran with the story they were handed.

Twenty times closer to its star than Mercury is to the Sun, Kepler 10b speeds round its orbit in just 20.1 hours. At such proximity, the star's heat scorches the world to more than 1,500 degrees Celsius, meaning that the Sun-lit face is almost certainly covered by lava lakes. This too was a rerun of the CoRoT-7b press coverage from two years before, yet only a few astronomers were honest enough to remind reporters that this was actually just Kepler's first rocky planet, not *the first* rocky planet. One of those honest voices was Sara Seager, who was part of the Kepler team, and a co-author on the discovery paper, who put the record straight for *USA Today*.[37]

In the rush to make the discovery sound Earth-like by calling it rocky, NASA may have missed a trick. Although the planet's density was mentioned in the press release, this was not really elaborated on. The density was extreme at 8.8 grams per cubic centimetre, which is greater than the density of iron. The Earth, for comparison, is just 5.51 grams per cubic centimetre. So, was the planet really 'rocky' or was it actually a metal planet? An accompanying video hinted that, being so close to the star, the planet's rocks were possibly being melted clean away, to escape into space as a gas, rather like the way the ice on a comet vaporizes to form a beautiful tail.

Just a month later, the Kepler mission team revealed that they were sitting on 1,100 possible planets in their data. It was more than double the number that had been found by all the teams in the world in the previous 16 years. To underline the achievement, the 1,100 candidates had come from only the first four months' worth of data, taken between May and September 2009. The rest of the data from the 156,000 stars in Kepler's field of view still had to be analysed. And the spacecraft was continuing to take more data, looking for planets with bigger and bigger orbits.

The team called these planetary candidates because they wanted to wait until radial-velocity measurements from the ground had shown the parent star's wobbling with the same period as calculated from the transits, corroborating the discovery. Borucki said at the time that, although they were being cautious about claiming true detections, he felt certain that the majority of the candidates would be confirmed in the coming months and years. Fifty-four of the candidates attracted immediate attention because they were in the 'habitable zone' of their parent star, and of these five were roughly about the size of the Earth. But what did the phrase 'habitable zone' actually mean? Indeed, what made a planet habitable? To find out, we have to look back to scientific work that began in the aftermath of the Second World War.

The person who pioneered the definition of what makes a planet habitable was probably a war criminal. Hubertus Strughold was a German physician spirited to America at the end of the Second World War. He was one of more than 1,500 German scientists, technicians and engineers identified as being useful to the United States, and so transported to live there, as part of Operation Paperclip.

In 1948, he became physician at the US Air Force School of Aviation Medicine at Randolph Field, Texas, and began to pioneer what he called 'space medicine', the study of the medical challenges posed by space flight. One of his projects was to build a space cabin simulator in which he could test human subjects. Strughold's dark secret was that he probably had more experience than anyone around him realized about the effects of extreme environments on the human body.

In the lead-up to the war, he had led the German air force's medical research unit, investigating the effects of altitude on human consciousness. When hostilities broke out, his team was absorbed into the Luftwaffe and heavily implicated in the Nazi human experiments, particularly those that saw prisoners from the Dachau concentration camp subjected to freezing conditions in order to test various methods of revival.

Despite being named by the Nuremberg Trials as being

implicated in these war crimes, Strughold was never brought to trial because numerous investigations seemed to stall during the collection of evidence.

In 1953, he wrote *The Green and the Red Planet: A Physiological Study of the Possibility of Life on Mars.* In the book, he discussed the conditions that a human needs to stay alive. He coined the term 'ecosphere' to mean the zone around a star that would provide enough light and warmth for liquid water to flow on the surface. Others researchers were coming to the same conclusion. Harvard astronomer Harlow Shapely published almost identical ideas that year, referring to a Liquid Water Belt around stars.

In 1964, Stephen H. Dole of the RAND Corporation, a not-for-profit think tank started in the United States after the Second World War, published a book called *Habitable Planets for Man.* In the preface he wrote: 'The space age is still very much in its infancy. To attempt at this early date to predict the ultimate future of space flight and its impact on human affairs would be like trying to forecast the complete career of a child barely out of the cradle.' Nevertheless, he went on to say that habitable planets around other stars were the ultimate destination of the world's burgeoning space programmes.

Using Strughold's definition of an ecosphere, Dole discussed what else determined whether a planet could be habitable. He listed seven criteria.

Firstly, the mass must be greater than 0.4 Earth masses, he wrote, in order for the planet to generate enough gravity to hold on to a breathable atmosphere, rather than have those gases escape into space. However, the planet could not be too big. It must contain less than 2.35 Earth masses so that the surface gravity is less than 1.5 times the Earth's pull. Dole thought this a prerequisite for any life to be easily mobile, and to be able to construct buildings.

Secondly, he listed the rotation rate of the planet as needing to be less than 96 hours. If it were any slower, he thought, the daytime temperatures would soar too much and the night-time temperatures would plunge too low.

Thirdly, he thought the age of the planet needed to be at least 3 billion years to give enough time for complicated organisms and a breathable atmosphere to develop.

His fourth criterion concerned the angle of the planet's equator, which gives the planet its seasons. Dole reasoned that if the equator was tilted too sharply above or below the star, then the level of illumination received from the star would vary over too a wide range throughout the planet's year. Hence the seasons would swing between extremes, perhaps ruining the habitability.

Number five on the list was the planet's orbital eccentricity, which measured how elongated was the orbit. This too would impose seasons on the planet because an elliptical orbit guides the planet closer to the parent star and then further away again. An elliptical orbit that was too large, Dole wrote, 'would produce unacceptably extreme temperature patterns on the planetary surface.'

The final two criteria were about the central star. He thought it needed to be similar in size to the Sun because if it were too large, it would use up its fuel too quickly and stop shining before reaching the 3 billion years age he had assumed was needed for the development of life. If it were too small, then the planet would have to be much closer to receive the same amount of heat as Earth, and this would lock the planet into showing the same face to the star, violating his four-day rotation criterion.

Dole concluded that if all these requisites are satisfied, then there is a very good possibility that a planet will be habitable. Effectively he was describing Earth's twin, using just a bit of wiggle room around the basic parameters. Using these and other assumptions, he made a rough estimate that some 600 million habitable worlds would be found around the Galaxy's 100 billion stars. Although this is a huge number, it meant that the frequency of Earth-like worlds was less than 1 per cent.

In the years that followed, astronomers focused almost exclusively on Strughold and Dole's concept of the ecosphere: in other words, the size of the orbit. Using simple assumptions about the transparency of Earth's atmosphere and the reflectivity of the surface, Dole estimated the range of habitable orbits around a Sun-like star to be 0.725–1.24 astronomical units. A few others followed suit and, by the 1970s, the restricted range of possible orbits was being referred to as the Goldilocks zone because the planet would be neither too hot nor too cold. The term 'ecosphere' became superseded by the modern term 'habitable zone'.

This was the parameter that the exoplanet hunters focused on during the headline wars. The closer a planet got to the habitable zone, and the closer its mass got to our world, the more 'Earth-like' they said it was.

One particularly spectacular discovery was around the Sun-like star that the team had named Kepler 11. A complicated series of transit signals had resolved into a beautiful six-planet system. Five of these worlds orbited closer to their star than Mercury does to the Sun, while the sixth sat between the distance of Mercury and Venus. It was a revelation that planets could be so closely packed, and opened a route to estimating their masses without the need for radial-velocity measurements.

Back in the 17th century, Isaac Newton had used his newly developed theory of gravity to explain why Jupiter and Saturn pulled each other slightly off their normal orbits when they were close together. Kepler team member Jack Lissauer, from NASA Ames Research Center, and collaborators applied the same analysis to the planets around Kepler 11. Although the planets were all smaller than Jupiter and Saturn, they were much closer together and so their gravitational fields could readily affect each other. As the planets pulled each other off course, this manifested itself as small variations in the time at which the transits took place. Using these timing variations, the Kepler team found that they could calculate the masses of the planets.

Perhaps surprisingly, all of the planets were low-mass, low-density places, implying that they were rarefied, gaseous worlds. If so, they were far from gas giants. The smallest had masses just twice that of Earth. Whereas iron-world Kepler 10b was one of the densest planets yet found, Kepler 11's retinue were some of the least substantial planets known. Within a single month, the Kepler mission had greatly expanded the boundaries of what we thought was possible for planets.

Yet for all of this wonder – and they were wonderful discoveries – a lingering disappointment remained because none of these worlds was Earth's twin; they weren't even potentially habitable. Then, in December 2010, it looked like dreams were coming true.

*

The Kepler team were numbering their discoveries in order. Now it was the turn of Kepler-22b, and it was looking very good indeed. It was the first of the 54 habitable planet candidates that the Kepler mission had announced in February to be confirmed. The star itself was a G-type star, not a red dwarf, and the planet was in an almost Earth-sized orbit, circling once every 290 days, which gave it an orbital diameter 85 per cent that of Earth's. This was definitely looking good. The only downer was that the planet was 2.4 times larger than Earth. Nevertheless, it was without doubt the smallest of the exoplanets to be found in a star's habitable zone, and what's more it was a Sun-like star's habitable zone.

Borucki said that fortune had smiled upon the Kepler team because the first transit had been captured on 15 May 2009, just three days after the spacecraft was declared operational. The second transit had come on 1 March 2010. This gave the astronomers the orbital period and allowed them to make a prediction about when the third transit would happen if this really were a planet. On cue, it took place on 15 December 2010, and the astronomers were sure they had something. So when ground-based observations failed to extract a reliable signal to verify the planet independently and give the mass, they turned to NASA's infrared space observatory, Spitzer. As expected, the next transit took place on 1 October 2011, and was indeed seen by Spitzer. The planet was definitely there, now all they had to do was estimate its mass.

Although 2.4 times larger than Earth does not sound very much, it makes a big difference to the mass of the planet. Estimates suggested that it could be up to 100 times the mass of the Earth. If so, it would generate so much gravity that it would hold a much higher proportion of gas and ices than the Earth. Being in the habitable zone, this ice would probably melt and cover the planet in a global ocean. While there may be life in the ocean, the chances of it developing fire, and thus technology, seemed to be severely curtailed.

Kepler-22b was announced at the beginning of a conference dedicated solely to discussing the Kepler mission and its findings. The mission team had another 1,094 planetary candidates to follow up, bringing the total to 2,326. Of these, 207 were about Earth's size, 680 were super-Earths, 1,181 were Neptune-sized, 203 were Jupiter-sized, and 55 were larger than Jupiter. Of these worlds, 48 were now in what the team thought of as the parent star's habitable zone. The number had dropped because the team had got tougher at considering what the effects of a planet's atmosphere might be.

On Earth, the atmosphere provides a greenhouse effect which traps the Sun's heat and raises the temperature of the surface. Without this, our world's surface temperature would be around –18 degrees Celsius instead of the clement average of 15 degrees Celsius that we experience. Denser atmospheres give bigger greenhouse effects. Venus would be around 34 degrees but its choking blankets of gas boost it hugely to more than 450 degrees.

As the Kepler team thought these statistics through, they realized that Kepler-22b with its great mass would probably have a dense atmosphere, and so a huge greenhouse effect. If so, Kepler-22b could be a steamy, fog-ridden planet with no real definition between the ocean and the atmosphere. The water there would be in a constant state of evaporation and precipitation.

It seemed Nature was teasing us: we could find planets the same size as Earth, we could find planets in the habitable zone of Sun-like stars, but we could not seem to find Earth-sized planets in the habitable zone of Sun-like stars. Maybe such Earth analogues were not out there after all? Perhaps Earth's twin simply does not exist?

A typically confident Sara Seager was quick to scotch that idea. In 2012 she said, 'Earth's twin world is sitting in Kepler's data set. We haven't found it yet because we haven't completed the analyses.'

But, behind the scenes, the real reason for the lack of success was becoming apparent, and it was not good news at all.

NINE

Disaster

While the exoplanet hunters kept tight-lipped on the subject to the outside world, behind the scenes they were struggling. Disturbing discoveries were threatening to derail the search for Earth's twin.

ESA had originally been interested in a mission to study starquakes and other brightness variations before Fridlund had pointed out that such instrumentation could also reveal transiting planets. While it was indeed the exoplanet discoveries that were grabbing the headlines from the CoRoT mission, the stellar seismology work continued apace. There were 25 scientists looking for exoplanets but twice that number analysing the behaviour of the stellar surfaces. Completely unexpectedly they were seeing that G-type stars fluctuated much more than anyone had expected.

The churning gases found inside the Sun act like dynamos generating magnetic fields that rise up and burst through its surface, cooling the gas there and making sunspots – dark patches that can persist for weeks or months. These dim the star a little. At the other end of the scale, bright regions known as faculae and plages show up on the outer fringes of these magnetically active regions. The comings and goings of these magnetically driven fluctuations affect the overall brightness of the Sun, and are referred to as stellar activity.

CoRoT's data showed that apparently Sun-like stars were up to twice as variable as our own and this made it far more difficult to pick out the tiny drop in starlight created by an Earth-like planet. It was rather like the difference in spotting a tiny boat on a bobbing ocean to glimpsing it between huge waves. Although CoRoT could have seen an Earth-like planet around a perfectly Sun-like star, this unanticipated intrinsic variability placed Earth's twin out of their reach. All eyes turned to Kepler.

Over at NASA, scientists were seeing the same readings and coming to the same conclusion. It didn't kill their quest but it was a wounding blow. In deciding how big to make the telescope they had assumed that all Sun-like stars would display Sun-like activity. Now they were seeing that the Sun was particularly quiet; it might even be in the quietest low percentage of otherwise identical stars.

As the Kepler scientists looked into what this meant for them to meet their objective of determining the frequency of Earth analogues, they calculated that, instead of their original estimate of three years to find Earth's twin, now it would take between five and eight years to collect enough data so that the planet's regular, repeating signal could be distinguished from the random ups and downs in brightness.

In April 2012, when Kepler reached the end of its primary mission, NASA granted the astronomers an extension. It would fund the mission for at least another two and a half years. That would give the team five and a half years of data, and then another decision could be made about extending to the full eight years. Through their Twitter account

(@NASAKepler), the team said that they were grateful and ecstatic, but the latter feeling was short lived.

Just a few months later on 16 July, Kepler transmitted bad news: one of its reaction wheels had failed and the spacecraft had shut down. This was the nightmare that Bell Aerospace and NASA had anticipated in the two years preceding launch, and worked so hard to avoid.

After a few days to fully diagnose the severity of the problem, the engineers realized that the faulty reaction wheel was beyond revival. So they commanded Kepler to begin working again, employing the spare reaction wheel. For good measure, they also announced another 500 planetary candidates, bringing their tally to 2,800.

The mission was back to full operation but there was clearly concern about how quickly the third reaction wheel had failed; the team had hoped it would last longer than three and a half years. If another one went, the search for Earth's twin would end with it. The team needed four more years, and now, without a back-up, they were on borrowed time.

As those at NASA got back down to routine operation, a geologist turned astrobiologist from Washington State University set about defining mathematical ways to quantify 'Earth-like' and habitability.[38] Dirk Schulze-Makuch was born and educated in Germany but moved to America before obtaining his Ph.D. in the geosciences. Soon afterwards, he became interested in assessing the habitability of worlds within our solar system, and now he was ready to rally colleagues and think about the exoplanets. He was mindful of the need to stop the rivalry about claims to have found the 'most Earth-like' planet. What did that even mean from a scientific point of view? Surely, thought

Schulze-Makuch, there was a numerical way to quantify the concept.

The first approach he took was the Earth Similarity Index, which compares an exoplanet's radius, density, surface gravity and estimated surface temperature to Earth's. Based on these physical properties, it gives a number between zero and 1, with 1 being a planet having all four properties identical to Earth. Although it is not exactly correlated to habitability, the nearer to 1 an exoplanet scores, the more Earth-like its conditions and, since Earth is clearly habitable, this could be taken as a clue that the planet might be habitable.

Next he asked himself, 'Does an exoplanet have to be Earth-like to be habitable?' He and colleagues were trying to move away from the bias that Earth was the only model for a habitable world, and reasoned that all that was needed were similar conditions to Earth, rather than similar physical properties. So they looked more widely at what might be needed to make a planet habitable to microbes, and defined a second measure: the Planetary Habitability Index. This included whether a planet had a rocky or a frozen surface, whether it had an atmosphere and a magnetic field. Other conditions were whether there was enough energy either from sunlight or tidal forces to power life, whether there were organic compounds present on the planet in sufficient abundance, and whether there was a liquid available that may act as a solution in which biological reactions could take place. The scale ran again from zero to 1 but this time it was not scaled to the Earth. The closer to 1, the more probable it was that life could be present.

While the Earth Similarity Index (ESI) was designed to be applied to the data that could easily be collected about exoplanets, the Planetary Habitability Index (PHI) required much more sophisticated analysis that would only be fully possible if the Darwin/Terrestrial Planet Finder missions were resurrected. To show the difference between the two indices, Schulze-Makuch and colleagues calculated the values for various planets and moons within the solar system.

The ESI scores came out in order as Earth (1.00), Mars (0.70), Mercury (0.60), Venus (0.44), Io (0.36), Callisto (0.34), Ganymede (0.29), Ceres (0.27), Europa (0.26), Titan (0.24). Apart from the planets, this list also includes moons of Jupiter (Io, Europa, Ganymede and Callisto), Saturn (Titan); and the largest asteroid Ceres. When they switched to the PHI, however, things looked very different: Earth (0.96), Titan (0.64), Mars (0.59), Europa (0.49), Jupiter (0.37), Saturn (0.37), Venus (0.37). Now, worlds that were in no way Earth-like, Jupiter and Saturn for example, scored higher than worlds that were more 'Earth-like' in terms of their bulk properties.

For comparison, the ESI scores of some of the more prominent exoplanets at the time were (in reverse order): Gl 581 d (0.53), Kepler-22b (0.71), the disputed Gl 581 g (0.76), and the clear winner Gl 667C c (0.84), which was still mired in an argument over who had found it first. Yet, as the PHI showed, the question of habitability was perhaps not so closely tied to how similar the world was to Earth's bulk properties. In other words, finding Earth's twin was not necessarily a prerequisite for finding a habitable planet. But when it would be possible to gather the data necessary to perform PHI calculations was anyone's guess, and some exoplanet hunters were beginning to feel frustrated at the narrow options for progress.

Spirits picked up in April 2013, when the Kepler-62 planetary system was made public. Here were five planets in orbit around the same star. Admittedly the star was considerably smaller and dimmer than the Sun, but planets e and f sat in the habitable zone. Kepler-62e was 1.6 times the diameter of the Earth and received 1.2 times the amount of solar energy, while Kepler-62f was just 1.4 times the size and received 0.41 times the quantity of Earth's illumination. They orbited their star in 122.4 and 267.3 days respectively. Astronomers thought both were possibly habitable, with e scoring 0.83 on the ESI scale, and f scoring 0.67.

Hot on their heels, a team involving Marcy announced Kepler-69c, with a radius of 1.7 Earths. This planet orbited close to the habitable zone of a Sun-like star, and although it was not really better than the previous finds, the team felt inspired enough to say that it represented 'an important step on the path to finding the first true Earth analogue'. Within a month, however, more detailed calculations had revealed that the planet was probably more like Venus than Earth, and they rolled back on the idea that it was habitable.

These discoveries, and the general sense of anticipation that an Earth analogue must be on its way, led to a joint congressional hearing in the United States on behalf of the US House of Representatives subcommittee on space, and the subcommittee on science and technology. The question to be addressed by the three expert witnesses was 'Have we found other Earths?'

Laurance Doyle, a Kepler scientist from the SETI Institute, Mountain View, California, spoke first. By way of context he described his own discovery: an intriguing planet that orbited two stars not one. Called Kepler-16b, it was a gas giant that held a stable 229-day orbit around the pair of stars. Although the planet itself was unlikely to be habitable, the fact that it orbited twin suns led Doyle and his team members to remember the scene in *Star Wars* when a melancholic Luke Skywalker watches the double sunset on his home world of Tatooine.

Wanting to make reference to the film at the press conference, Doyle contacted the makers of *Star Wars* and asked for permission. Not only did the film's creator George Lucas say yes, he even sent one of the special effects team down to the press conference, who showed a clip of the movie and told journalists, 'Again and again we see that the science is stranger and weirder than fiction. The very existence of this discovery gives us cause to dream bigger.'

At the conclusion of his testimony, Doyle returned to whether we had found other Earths, 'The safe answer to the question is "almost". Within the next few years, Kepler will be able to discover exactly Earth-sized planets . . . In the next few years we will have the privilege of answering this age-old question: in the universe is there another place like home? I think with the Kepler mission we are just on the verge of answering yes.'

Former astronaut John Grunsfeld spoke next. Although his days in space were over, he had gone on to become the head of science at NASA. After outlining the agency's commitment to exoplanet research, he said that he was confident that it was not a question of whether or not they would find an Earth-like planet, but when. The final speaker was James Ulvestad of the National Science Foundation (NSF). He reminded the subcommittees that the NSF had consistently funded exoplanet research and was currently helping to develop the technology that would assist in the discovery of Earth's twin.

During the question and answer session, Grunsfeld spoke

of the eventual need for a mission that could look for the signatures of life around nearby Earth-like worlds, once these worlds had been identified. This was significant because he was, in effect, putting a replacement for TPF/Darwin back on the table. The message was clear: exoplanet astronomers were dusting themselves down after the disappointment of the decadal review. Momentum was gathering again, and Kepler was on course for finding the first example of Earth's twin planet within the next few years.

Then disaster struck.

There are simply too many spacecraft in orbit and not enough radio dishes to keep in contact with all of them at the same time. When a mission is planned, the amount of contact must be defined so that it can be scheduled. For Kepler, ground controllers spoke to the spacecraft just twice a week. During this time, they downloaded everything it had recorded in the last few days and also its telemetry, the 'housekeeping' data, which told controllers the condition of the craft and its components.

On 1 May 2013, the link-up showed that the spacecraft had shut itself down into 'safe mode'. This low-power mode is similar to sleep on a computer. It clicks in during emergencies, such as when the reaction wheel had failed in July 2012. It is designed to simply point the solar panels at the Sun, the spacecraft's antenna to Earth, and keep them there to await further instructions. With their hearts in their mouths, the team asked it to download all its telemetry so that they could start looking for the cause of the problem.

After a few days of analysis, the indications were that the star trackers were at fault. These are little telescopes that judge the movement of the spacecraft by recognizing the constellations. People breathed a collective sigh of relief that it wasn't the reaction wheels. They restarted the spacecraft and returned it to full operation.

A fortnight later, the spacecraft put itself back into safe mode again. This time something was clearly very wrong; instead of pointing fixedly at Earth, the spacecraft was rotating slowly. The main antenna was only intermittently pointing at our planet, and this limited the contact that ground controllers could have with the spacecraft. Nevertheless, during one of these communications windows, the ground crew ordered Kepler to stop rotating using the reaction wheels. Although the manoeuvre worked, there were clear signs in the telemetry that one of the remaining reaction wheels was wearing out. NASA relayed the glum news to the world on 15 May:[39] the spacecraft was down to two reaction wheels and so could not point accurately. All planet-hunting attempts were suspended. Attempts to compensate for the reaction wheel using thrusters were discussed, as were options to try to revive the faulty component, but the thruster option used too much fuel and the component proved unrecoverable. In August that year, NASA called it quits. The Kepler mission was over and so was the hope of finding Earth's twin, at least for the time being.

Marcy knew exactly the magnitude of what had been lost. He told *Sky and Telescope* magazine,[40] 'I am just devastated. My hands are trembling, and my heart is aching.'

With the data they already had, exoplanet researcher Sara Seager estimated that they would be able to find Earth-sized worlds in orbit around Sun-like stars with orbital periods up to 200 days long. This would reveal planets like Venus, which is the size of the Earth and in a 225-day orbit around the Sun. Although it was difficult to call Kepler a failure when there was so much valuable data to be analysed, it was clear that the spacecraft had failed in its primary objective of determining the frequency of Earth-like worlds, and this would have serious knock-on effects.

For instance, it makes designing a follow-on mission difficult because the size of the telescope needed depends upon how close habitable planets are likely to be to us. If Earth's twins are rare, any that we find are likely to be far away because they are likely to be spread few and far between. So the telescope needs to be larger and therefore more expensive. If Earth's twins are common, we are more likely to find one closer to us, and so the cost of the telescope can be kept down. Kepler had provided this statistic for potentially habitable planets around red dwarf stars, showing that on average one in six red dwarf stars has a rocky planet in its habitable zone. This showed that the nearest habitable planet could be within 15 light years of Earth, making it very close. But Sun-like stars are rarer than red dwarfs. Astronomers already knew that G-type stars like the Sun make up just 10 per cent of the stars in the Galaxy. Now they also knew that the Sun may be one of the small percentage of least active stars, making another truly Sun-like star extremely rare, perhaps only one in a thousand. And that star may or may not have an Earth analogue in orbit. All tallied up, Earth's twin could well be many tens or even hundreds of light years away. With the end of the Kepler mission, we simply have no way of knowing.

It wasn't just the stars that were causing concern. The exoplanets were also making astronomers question their basic assumptions: chiefly that most exoplanet systems would follow the outline of our own. In other words, small rocky planets near the star, followed by large gaseous planets further out. The truth was that the majority of planets that were being revealed simply did not conform to this pattern.

Nature, it seemed, like to mix the sizes and generally compress the orbits. There are now about 500 stars known to be home to more than two planets. Of these, only Kepler-90 shows any semblance of similarity to our own planetary system. It is similar to the Sun in terms of size and brightness, and in the autumn of 2013 it was found to be home to seven transiting planets.

Located at 2,500 light years away, the system was too far for radial-velocity searches to confirm and measure the masses of the individual planets, so all the astronomers had to go on was the sizes of the planets, but this was enough to show the similarity. The first two worlds were just a little larger than Earth at 1.31 and 1.18 times the Earth's diameter. The next three worlds were super-Earths at 2.88, 2.67 and 2.89 times the Earth's diameter. The last two worlds were giants at 8.13 and 11.32 Earth's diameter. So the pattern was clearly one of smaller, probably rocky planets followed by larger, almost certainly gas giant worlds.

When it came to the orbits, however, the difference to our own planetary system was only too apparent. The Earth-sized planets completed an orbit every 7 and 8.7 days. The super-Earths took 60, 92, 125 days to orbit, and the gas giants took 211 and 332 days. All seven planets were crammed into an area smaller than the Earth's 365-day orbit.

Other multi-planet systems were more higgledy-piggledy. The HARPS team had found an extraordinarily confusing radial-velocity signature for HD 10180. This was a clear sign that there was a multitude of planets all pulling the star in different directions at different times. It made extricating the individual planets difficult but the team eventually settled on a seven-planet system. Although the general trend was for the largest planets to be the outermost, the arrangement of the inner worlds was more confusing with minimum masses for the planets running 1.3, 13, 11.9, 25.0, 23.9, 21.4 and 65.8 times the mass of the Earth. For reference, Uranus is 14.5 Earth masses, and Saturn is 95.2. All were contained within a distance compatible to Mars' orbit.

What was becoming clear was that there were a lot of multi-planet systems out there but very few that looked like our own. Most had their planets much more tightly packed together and this growing realization led some to doubt the prevalence of Earth analogues.

Marcy attacked the problem from the other end. He knew that finding Earth's twin from the ground was technologically impossible with the equipment they had, but finding Jupiter's twin planet around a Sun-like star should just be a waiting game. If they could find Jupiters, this would be suggestive of planetary systems with the same architecture as our own – small rocky planets followed by large gaseous ones. The researchers would joke that finding Jupiter analogues meant that there was plenty of space left for Earths in between them and the star.

Clearly, researchers were going to find the shorter-period planets first because they required less time to watch them complete their orbits. As time had gone by, planets in larger orbits had been increasingly reported. Marcy's group found gas giants with masses of 1.07, 2.2 and 2.96 times that of Jupiter, which orbited their more-or-less Sun-like stars in

119, 445, 230 days. Yet, this was still a long way off from the 12-year (4,380 day) orbital period of Jupiter. As it would take at least an orbital-period's worth of data to see what was happening, then another half an orbital to be absolutely sure that the cycle was repeating, finding Jupiter's twin was going to take the best part of two decades. Such a survey had been Marcy's intention all those years back, when he and Paul Butler had set up their planet search in the first place.

Now it was 2010, Marcy and his new collaborators reported the results[41] of the long-term monitoring campaign that he had started in 1997. Called the Californian Planet Survey, there was one planet that looked pretty similar to Jupiter. It was HD 13931 b. Admittedly it was at least 1.88 times the mass of Jupiter but it was in an almost circular orbit around a reasonably Sun-like star, and took 11.5 years (give or take a year) to complete an orbit. But as encouraging as this initially sounded, few others were turning up.

The Anglo-Australian Planet Search, manned by an independent team of exoplanet hunters, reported that, after taking data for 12 years, they estimated only 3.3 per cent of stars in their survey showed signs of a Jupiter analogue.[42]

It seemed that, however astronomers came at this problem, they were running into the same conclusion. Earth's twins may not be common after all because our solar system may not be that similar to others.

Clearly, another survey was needed to fill in the gaps in our knowledge. This survey should look at nearby stars because those would allow more detailed follow-up in the near future. The good news was that Seager and others were already hard at work on such a mission.

*

Sara Seager has always loved astronomy; some of her earliest memories are of the Moon. Growing up in the street-lit environs of Toronto, she remembers being shocked by the sheer number of stars in the night sky when she was away from the city lights. At high school she discovered how appealing physics was because whole concepts could be encapsulated in equations. One of the best days of her life was the discovery at a university open day that you could become an astronomer.

By the mid 1990s, Seager had made it to graduate school at the Harvard College Observatory. Back then, the place had been electrified by the discovery of exoplanets and, unlike many other astronomers, who were taking a sceptical line, those at Harvard believed the work. Her Ph.D. supervisor encouraged her to work on theoretical predictions of what the atmospheres of hot-Jupiter planets would be like. She developed computer models of how Jupiter's atmosphere would react under the influence of so much heating close to the star and found that such planets could well have atmospheres of water vapour and methane, together with vaporized sodium and potassium. It was the kind of result that only another exoplanet researcher could fully appreciate but it made her early reputation.

After gaining her Ph.D., she moved to the Institute for Advanced Study, Princeton, New Jersey. There she was inspired to continue her exoplanet work by Professor John Bahcall, who had supervised the development of the Hubble Space Telescope. He told her that any avenue of enquiry was valid, so long as it was rooted in sound physics, and the phenomenon itself was likely to be detectable (and therefore tested) in her lifetime.

It was the idea that her work should be testable in her lifetime that was now starting to haunt her. By the time she reached her early 40s, the loss of Darwin/TPF made her realize that the missions that would help test her work were not guaranteed to happen. She could be devoting her life to things that she would never know were valid or not. This feeling of the clock ticking, and having to make things happen, was part of the motivation for her organizing the 2011 one-day conference 'The Next 40 Years of Exoplanets' where the assembled astronomers had picked over the bones of TPF's cancellation.

Back then, she had declared plaintively, 'I want to be part of the people who find the planets, so that thousands of years from now, when people are embarking on their interstellar journey, they will look back and remember us as the people who found the planets like Earth around the very nearest stars.'

The plain truth was that, even if Kepler had discovered Earth's twin, it would have been too far away for astronomers to do anything about it. Kepler deliberately looked at a single distant star field, so that it could see a great many stars all at once, but that meant that all of its planetary discoveries were so distant that it put them out of reach for the follow-up analysis that would reveal their atmospheres. The reflected light from the planet would simply be too dim by the time it had crossed all that space to reach us. Even measuring the exoplanet's mass with radial-velocity measurements from the Earth was proving too challenging in a growing number of cases.

Even when Kepler was functioning and on course to find Earth's twin, Seager and others were aware that a follow-on survey would probably be needed to pinpoint the stars closest to Earth that hosted planets. Yet she was also becoming disillusioned with the enormous timescales and high price tags that seemed to go hand-in-hand whenever NASA became involved in a mission. So she was looking at the extreme miniaturization of technology that was promising to revolutionize access to space for the private sector.

Back in 1999, professors from California Polytechnic State University and Stanford University had developed the blueprint for a 10cm×10cm×10cm 'CubeSat'. It was designed to be built by graduate students to teach them the essentials of satellite design and construction, and it was proposed to have the same capabilities as the world's first satellite Sputnik 1, which was a 58-centimetre-diameter sphere that broadcast a radio signal to Earth in 1957. Almost immediately engineers and scientists began to see other uses for CubeSats, especially if two or three were joined together.

With her students, Seager looked at whether it was possible to build miniature space telescopes. By the time of the one-day conference she had a design and a mock-up to show the audience. The idea was that, instead of building a single spacecraft to do everything, you build many CubeSats to do all the different things a traditional spacecraft would do: one would contain the telescope to look for transits, another would handle the communications with Earth. The CubeSats would be so cheap to manufacture that, instead of using the same telescope to look at many stars, you could construct many telescopes with each one dedicated to looking at a particular star only. If one telescope were to fail, instead of losing the whole mission, you would have the option to launch a spare or manufacture and launch a new one. She called the concept ExoplanetSat. With an estimated cost for the first telescope at

$5 million, it was peanuts for a space mission (Kepler had cost $600 million). And, once the miniature telescope production line was up and running, the cost would tumble further. It was a new way of working, and Seager was hawking the idea to private companies interested in developing cheaper ways to access space.

Simultaneously, at MIT, there was another exoplanet mission idea in the pipeline. Seager was part of the science team but was not in the driving seat. Instead, the principal investigator's role was taken by Senior Research Scientist George Ricker, who had a 40-year career at MIT masterminding space missions. His idea was the Transiting Exoplanet Survey Satellite (TESS), designed to do essentially the same as ExoplanetSat: survey the nearest stars, looking for transiting planets down to Earth's size but without the emphasis on finding Earth's twin. Instead, this mission took its inspiration from the work of Schulze-Makuch in defining a habitable planet.

It suggested that the easiest route to finding a habitable planet was not to look for Earth's twin but to look for Earth's cousin. Instead of concentrating on Sun-like stars, the rationale was to survey the nearest stars, which were primarily red dwarf stars, looking for planets in the habitable zone. In the local neighbourhood of stars, the spherical region of space out to 30 light years, there are no O and B-type stars, 4 A stars, 6 F stars, 20 G stars (Sun-like stars), 44 K stars, 246 M stars, and 20 dead stars called white dwarfs. This is in line with expectations that low-mass, red dwarf stars (K and M stars) form more easily than high-mass stars (O and B). So there are more than ten times as many red dwarf stars within reach as there are Sun-like stars.

Using four 10-centimetre telescopes, TESS would target

different areas of sky for a month or so at a time looking for transits. It would eventually cover most of the sky and in so doing would discover the nearest examples of rocky planets in the habitable zones of these red dwarf stars. These would become targets for the James Webb Space Telescope to study further, revealing their atmospheric compositions after its launch in 2018.

Ricker had been developing the idea of TESS since 2006 when he too had thought that private funding might carry the day. Indeed, he attracted private development funding from Google, the Kavli Foundation and MIT donors. However, by coincidence, NASA had formed the Small Spacecraft Division at their AMES Research Center, which is located at Moffett Federal Airfield in California's Silicon Valley, in the same year. As space missions go, TESS is straightforward and so attracted the new division's attention straight away. A year later Ricker set his sights on it becoming one of NASA's Small Explorer-class missions. In 2008, the proposal was selected for an initial NASA-funded study and it won more study funding three years later.

Then in April 2013, a month before Kepler gave up the ghost, NASA announced that it had selected TESS for implementation from 11 different mission proposals that covered the entire field of astronomy. Cost capped at $200 million, the spacecraft was scheduled for launch in 2017 but could slip into 2018.

On 4 December 2013, soon after the demise of Kepler, the House Committee on Science, Space and Technology called another hearing to discuss exoplanets. This time it was entitled 'Astrobiology: Search for Biosignatures in our Solar System and Beyond' and one of the speakers was Sara Seager.

She used her allotted time to argue that humanity stands on a 'great threshold of space exploration. On one side of the threshold we now know that small planets exist and are common. But on the other side lies the possibility to find the true Earths, with signs of life,' she said, 'The point I want to make is this is the first time in human history we have the technological reach to cross the great threshold.'[43]

The unmistakable burgeoning of interest in the search for life on other worlds was highlighted by the chairman, Lamar Smith, who gave the statistic that in 1995, when the first exoplanet was discovered, fewer than 50 scientific papers were published that year on astrobiology. By 2012 more than 500 papers were being published every year. But to turn this into tangible, observational science was going to take a bold new space mission.

At the beginning of the year, Seager had quietly been appointed chair of a NASA science and technology definition team to study such a spacecraft. It was the resurrection of the science goals of Darwin/TPF but, instead of a fleet of telescopes, it was going to use a 'star shade' to directly image and study Earth-sized worlds. This was a different way again from the interferometer or coronagraph that had been envisioned a number of years earlier.

The star shade would be a large circular structure, tens of metres in diameter, that would fly in tight formation in front of a space telescope and arrange itself to block out the blinding light from a central star. In this way the space telescope could look to see the dim light from planets peeping round the edge of the star shade. It would target already known exoplanets with the aim of analysing the composition of their atmospheres.

A second NASA study would resurrect the coronagraph, which placed the eclipsing mechanism inside the telescope. Both missions, should they be implemented, were to cost less than $1 billion.

Learning from previous mistakes, NASA began these studies so that there was time to reach consensus well before the 2020 decadal review began. Indeed, the final studies were released in March 2015 and the teams are now working to explain what they could achieve if the $1 billion ceiling were removed. They are hopeful that, by 2020, the National Academy of Sciences will recognize the importance of such a mission and reinstate it in the decadal review.

In Europe too, there is the realization that a follow-on mission is essential to maintain the science programme started by CoRoT, and also to expand the extraordinary work of the HARPS instrument. In 2012, before the drama with Kepler, the European Space Agency had committed to a dedicated exoplanet space telescope in collaboration with the Swiss Space Office.

It was chosen from 26 proposals that had been submitted in March, in response to a call for small mission ideas. Aware of the decade-long timeframe for developing and launching most missions, ESA had instituted a new small mission programme. Each mission was designed to answer specific questions that had arisen in the scientific community, and which could be nailed with a single, timely, cost-efficient spacecraft.

CHEOPS was perfect. The name stands for CHaracterizing ExOPlanets Satellite and the spacecraft will target stars that are already known to host planets. It will be the first space mission to do this for nearby stars, and its instruments will provide much more accurate measurements of the planetary radii than is possible at present. It will also discover additional, smaller planets in orbit around those stars, building up a better picture of what typical solar systems look like.

Planned for launch in 2017, the mission has a target lifetime of between 3.5 and 5 years. It could also follow up discoveries made by TESS, and will certainly help provide exoplanet targets for future study by the JWST, or large ground-based future facilities.

Yet, as exciting – even necessary – as these two missions are, neither TESS nor CHEOPS is designed to find Earth's twin. It is possible that they may stumble across it. But they are not designed to watch any star for the years it would take to be certain of a detection. For that, a dedicated expensive mission is still needed. The Americans seem to have lost interest, preferring to concentrate their efforts on red dwarf planets, and hoping for habitable ones there. But in Europe, the dream is still alive, thanks to Malcolm Fridlund, who was about to perform his next rise from the ashes.

This time his Phoenix looks destined to fly.

TEN

The end of the beginning

The Eddington mission had been taken very seriously by the European Space Agency. It had been cancelled in 2003 only because of budget troubles. Its predecessor, STARS, had lost out to the Planck mission by just one vote. Now ESA was in the middle of a fundamental rethink about what its long-term space science strategy should be, and this process led to the definition of a programme called Cosmic Vision. To start the process, ESA had asked for input from the various astronomers, planetary scientists and physicists at work in Europe.

It was a grand consultation on what should be the agency's priorities. In many ways it was similar to the decadal review in America that brought so much strife to the exoplanet community there in 2010.

The last such exercise by ESA had taken place in 1984 and resulted in the Horizon 2000 programme, and its extension Horizon 2000 plus. Now it was time for a whole new reckoning. The agency issued its call for theme ideas in April 2004 and over 151 papers flooded in – more than twice the number received in 1984.

These were collated and the most popular ideas were debated at an open workshop in Paris in September, with more than 400 scientists in attendance. More refinement took place by the internal ESA advisory committees – many of which are staffed by members of the academic community not employed by the agency – and a final public workshop was convened in April 2005 at the European Space Research and Technology Centre (ESTEC) in The Netherlands. A month later, the Science Programme Committee, the top advisory body to the ESA science directorate, approved the themes, and Cosmic Vision was born. The four questions that the agency gave priority to answering during the years 2015–25 were:

1: What are the conditions for planet formation and the emergence of life?
2: How does the solar system work?
3: What are the fundamental physical laws of the universe ?
4: How did the universe originate and what is it made of?

Admittedly they were broad topics. The real job was to turn them into specific missions. A call for proposals in March 2007 yielded an unprecedented 50 entries. Of these 19 were astrophysics missions, 12 concentrated on fundamental physics and 19 were for solar system exploration. One of the astrophysics proposals was submitted by Eddington's former project scientist, Fabio Favata. He sent in a proposal for PLAnetary Transits and Oscillations of stars (Plato), a heavily redesigned version of Eddington. The agency chose it as one of five to be studied for a medium-class mission, costing no more than half a billion euros, but Favata was no longer eligible to take on the study scientist role. He had moved from his office in the science corridor at ESTEC to ESA's headquarters in Paris where he was now the astronomy missions coordinator and would soon become the Head of the Coordination Office for the whole of the science programme. So the task fell to Malcolm Fridlund, who was still busy at work with CoRoT.

As the studies progressed, it became clear that there were just three missions that were truly capable of being ready. One was Plato. Another was Solar Orbiter, which was designed to go closer to the Sun than any previous mission. The third strong contender was a space telescope called Euclid that was designed to map galaxies across the universe . But they were chasing only two mission slots: M1 for launch in 2017 and M2 which would leave the launch pad in 2020.

When the technological readiness of these three missions was evaluated, all scored highly but Plato came out on top. If that had been the sole criterion for selection, the mission should have been chosen immediately, but other forces were at work. As the teams worked towards the 2010 selection, Plato suddenly had a fight on its hands.

Solar Orbiter was planned to be a collaboration with NASA. There was no real way for the agency to pull out, or even delay, at this stage. So the selection of Solar Orbiter was a virtual given for M1. That left the M2 slot, and a straight fight between Euclid and Plato. Both missions had strong support in France, where the national space agency (CNES) had been responsible for leading the CoRoT mission, and there was a clear appetite to continue with a bigger, follow-on mission through ESA. But astronomers at large were gripped by the enigma of dark energy, which Euclid would address.

The mystery had begun in 1997 when two independent

teams of astronomers discovered that the universe was apparently accelerating. It had been known since the late 1920s that the universe was expanding but all had assumed that the rate of expansion would be slowing with time. Using exploding stars as their yardsticks, two international teams of astronomers had worked throughout the 1990s to measure this deceleration only to discover completely the opposite was seemingly taking place.

There was no known energy or force that would drive this behaviour and so astronomers started to call it dark energy to underline its mysterious nature. In the decade that followed, the more astronomers studied the phenomenon, the more it underlined their ignorance, and many came to the conclusion that what they needed was a vast survey of the galaxies in space, as these would allow the acceleration to be mapped out more precisely than ever before.

So the choice was to look for Earth's twin planet with Plato, or try to pin down the mysterious dark energy with Euclid. Given the way dark energy captured the imaginations of many astronomers, Euclid was judged to be the more urgent mission and was award M2.

In the normal run of things, this would mean Plato was scrapped. If Fridlund wanted to try again, he and his collaborators would have to repropose the whole mission with a different name and significant changes. But the ESA advisors came to his aid this time. They mandated that Plato should be carried over to the next selection round that would take place in four years. This placed it in the very best possible situation for a shot at the M3 mission, launching in 2024. Then a disaster took place.

With Euclid in the bag, CNES dropped out of wanting to

mastermind Plato. If no one else could be found to replace it, the study would have to be withdrawn. Fridlund began to search and found that the German Space Agency, DLR, liked the mission. It undertook to sound out industrial and academic partners, and quickly built a new consortium of people and companies with the right experience to make the mission happen – should it be selected.

In 2014, Plato was again before ESA's selection committee. It faced competition from missions that wanted to study matter falling into black holes, return samples from an asteroid and probe for weaknesses in Einstein's General Theory of Relativity. It also had a direct competitor in a proposal called the Exoplanet Characterization Observatory (EChO) which aimed to analyse larger exoplanet atmospheres.

Given the heritage of the effort that had gone into Plato, which meant that its technological readiness was head and shoulders above the others, and the ambition it had of finding nearby Earth analogies, there was really only one choice the committee could make. Fridlund remembers that chairperson Catherine Cesarsky, an eminent French astronomer, told him that Plato had the most compelling scientific case she had ever seen.

In February 2014, Plato was officially given the green light and set on course for a launch from French Guiana atop a Russian Soyuz rocket in 2024. It will carry a cluster of 34 telescopes, most taking a picture every 25 seconds and two capable of recording an image every 2.5 seconds. Plato will have a field of view 20 times larger than Kepler's. It will measure planetary radii to better than 2 per cent accuracy, mass to better than 10 per cent and usually better than 5 per cent. With accuracies like that, computer models can start to put constraints on the size of the exoplanet's iron core and the depth of the atmosphere.

The key to doing this will be Plato's unparalleled ability to measure the movement of the stellar surfaces so that these signals can be subtracted from the planetary data. Fridlund is confident that Plato will allow us to do physics on these worlds that we can really trust. There will be none of the ambiguity and wiggle room that has dogged the field thus far, and which has led to the endless stream of 'most Earth-like planet for sure this time' kind of announcements.

With an ESA exoplanet mission finally in the bag, Fridlund decided it was time to leave the agency. He is now having a busy 'retirement', having been snapped up to be a professor at universities in Onsala, Sweden, and Leiden, The Netherlands. His goal now is to pass on as much knowledge as he can to a new generation of astronomers, and he will continue his research once he finally does decide to slow down. That new generation have a lot to look forward to because one thing is for sure: if Earth's twin planet is out there, Plato will find it sometime after its launch in 2024. Unless of course someone finds it first.

In July 2015, it looked as if exactly that had happened.

The story begins on 10 May 2014, more than twelve months before the public announcement, when NASA scientist Jon Jenkins received an unusual email. Jenkins was a veteran of the Kepler mission, having put in 20 years. Now he was in charge of overseeing new software that it was hoped would make it easier to find planets in the noisier than expected Kepler data.

Such a process is called data reduction. Just like a chef

reduces a sauce to drive off the excess moisture and concentrate the flavours, so an astronomer removes as many of the extraneous signals as possible to leave the data in its purest form. After the Kepler mission came to an end, the engineering teams started to look again for what they call 'systematic errors'. These are errors introduced into the data because of the behaviour of the spacecraft. If the engineers could recognize these errors, for example by noticing that a certain distortion appears only when a certain component is switched on, then they could write software to remove it. In this way, they hoped to be able to compensate for the worse-than-expected stellar activity that was clouding the data.

The lead scientific programmer on this endeavour was Joe Twicken and it was his name on the email. Jenkins had first met him in 1992 when they were both working on ways to study planetary atmospheres from a distance. He knew Twicken to be highly experienced, guarded in his language, and utterly trustworthy in his judgement.

The email stated that, in reviewing the results of a test run of the new software, Twicken had noticed that it had identified a planet of 1.1 Earth radii, which was clearly in the habitable zone of what appeared to be a perfectly Sun-like, G-type star. This was it. Earth's twin, exactly the kind of planet and star combination that Kepler had been designed to find. Jenkins clearly remembers the cocktail of feelings that the message evoked: 'It was an exciting, thrilling, but also terrifying moment.'

They began an investigation and almost immediately hit a bump in the road. Although the star had been assigned the same temperature as the Sun, its radius was listed as just 80 per cent. This was odd because the mass and therefore size of a star determine its temperature. If two stars are exactly the same temperature, then they should be the same size as one another.

It is crucial to get this right because the size of the star is used to infer the size of the planet from the transit data. The 1.1 Earth radius was based on the star being just 80 per cent the size of the Sun. If the star was more Sun-sized, then the planet's radius would be larger, and it would no longer be Earth's twin.

With so much at stake, Jenkins and Twicken contacted astronomers who had access to ground-based telescopes, and asked them to study the star to determine its size more accurately. The initial results were not encouraging. The first results said that it was indeed the same temperature as the Sun but now 20–25 per cent larger than the Sun, making the planet 1.8 Earth radii. At that size it was probably not a rocky planet but a gaseous one. Even if it was predominantly rocky, it would have such a thick atmosphere because of its greater gravitational field that most probably it would not be very similar to Earth at all.

Then new observations came in suggesting that the star was larger than the Sun but only by about 5 per cent. That brought the planet down to about 1.5 Earth radii, and although it could no longer be thought of as Earth's twin, at least it could legitimately be called akin to the Earth. Computer models based on previous discoveries showed that at 1.5 Earth radii the chances that it is rocky are high.

The certain way to check would have been to measure a radial velocity for the star and turn that into a mass, which could then be used to calculate the planet's average density. Unfortunately with an orbit of 385 days at 1.05 astronomical untis (5 per cent further from its star than Earth is from the Sun) the radial-velocity signal would be too small to measure from a ground-based telescope. Even if there was an instrument with the correct sensitivity, the team would have to observe the star for more than a year to see a complete orbit. As this was impossible, Jenkins and his team embarked on months of work to check if it was possible – however unlikely – that a combination of faint stars could be fooling them.

By October they were convinced that they were not in error and began putting their analysis in a paper for publication after suitable review by other independent astronomers. But they didn't have Earth's twin. The final radius they felt comfortable in claiming from their calculations was 1.6 Earth radii. They thought this made it likely that the planet was rocky, so somewhat Earth-like, and being so similar to Earth's distance from the Sun, they were in favour of thinking of it as a potentially habitable planet.

The trouble of course is that habitability depends not just on the star but also the planet, and its atmosphere's ability to trap heat. More massive planets, like this one, might have denser atmospheres and so trap more heat. Their calculations also suggested that the planet receives 10 per cent more radiation than the Earth does. Although both of these considerations were going in the wrong direction, they still felt justified in talking about a good chance of habitability.

By July 2015, all the work was done, the paper was in its final vetted form and NASA called a press conference. In collaboration with the NASA PR gurus, they settled on the epithet not of Earth's twin, but of Earth's cousin. It was also just a few months before the 20th anniversary of the discovery of 51 Pegasi b, the first of the exoplanets. Didier Queloz dialled into the press conference to say a few words about how far the field had come, and how much there was still to do.

Although in the end it wasn't Earth's twin, it was the closest analogue we have. The press and the public went wild for it.

Concurrent with this announcement came another, less well covered story. William Borucki, who had worked at NASA for 53 years, 32 of them on the Kepler mission with 17 of those before the mission was even approved, was to retire. It was a perfect time to end his career. The mission he had championed was effectively over and he had been awarded the Shaw Prize – the same prize that Geoff Marcy and Michel Mayor had been awarded a decade earlier for their work on exoplanets.

In a statement issued by NASA's headquarters in Washington, John Grunsfeld, head of NASA's Science Mission Directorate, said, 'Bill's unique leadership, vision, and research tenacity has and will continue to inspire scientists around the world. He retires on such a high note that he leaves a legacy of inquiry that will not only be celebrated, it will be remembered as opening a new chapter in the history of science and the human imagination.'

It was not until the summer of 2016 that the next big planetary discovery (or should that be small planetary discovery) really hit the headlines. This time it was seismic news; one for the history books. But it caused no sense of excitement for its discoverer. Instead, because of his past experiences in the exoplanet community, he felt only trepidation when he saw the signal emerge from the data.

Guillem Anglada-Escudé, of Queen Mary University, London is the astronomer who had developed a new method of analysing exoplanet data. He had been drawn towards the search for other worlds partly because of his interest in science fiction, and his feeling that looking for planets around the stars encapsulated a sense of exploration.

His scientific acumen had allowed him to spot that the analysis software being used by the exoplanet hunters was specifically designed to look for planets around Sun-like stars. Now that the searches were moving to smaller stars, the software needed to be updated because the cooler stars presented a different set of signals than the Sun-like stars.

As soon as he completed his software, he tested it on publicly available data that he thought had been discarded by the other teams, and began to find unreported planetary signals almost at once. The most promising of these was the potentially habitable planet around red dwarf star Gliese 667C but he was scooped to the announcement (see chapter 8). Nevertheless, he went ahead and published his own report on the planet just after the original team and an angry exchange over who had made the discovery first began.

The very astronomers whose ranks he had hoped to join seemed to turn on him. He says it made him furious and he hit back; he would not be stopped. There were hundreds of stars in the public archives and he intended to look at them, and then collect more data to verify the planetary candidates his software spat out.

It was no easy task; rightly or wrongly he felt that the controversy over Gliese 667C was being held against him. His requests for observing time and for grant money were being refused, his results papers were being given a rougher than usual passage into print. As the fighting went on, it started to make him physically ill and he decided that he had to make efforts to heal the rifts.

The chance came in July 2015, when Debra Fischer, an exoplanet astronomer who has moved from San Francisco State University to become professor at Yale University, Connecticut, organised a workshop for those in the field. It was entitled 'Extreme Precision Radial Velocities' and as Anglada-Escudé remembers, it was an opportunity for everyone to come together and say some things face-to-face to clear the air. It wasn't just Gliese 667C they needed to talk about: there were other gripes between other teams that also needed to be resolved.

The workshop also presented a way for Anglada-Escudé to demonstrate once and for all the power of his software. By now he had teamed up with Mikko Tuomi, University of Hertfordshire, UK, who had developed a precise statistical way to combine disparate data sets so that all observations of an individual star could be used to make even smaller planetary signals stand out. At Yale, the participants had agreed to take part in a simple challenge. Fischer had organised for 30 sets of test data to be generated in which planetary signals were buried: the attending teams should bring their software and see who could find the most planets. Nine teams took part.

At the end of the test, the results were conclusive. Anglada-Escudé and Tuomi found the most planets and generated the fewest false alarms. It was a magic moment for them, and one that really saw their fortunes turn around. Concurrent with the workshop, they were informed that their latest proposal to use the HARPS spectrometer had been accepted. And with that, they knew that they had been accepted.

Their HARPS project was not just a shot in the dark. They were after a very special planet indeed. They had seen hints of it in the archival data and wanted to confirm or refute its existence.

The planet was about 1.3 times the Earth's mass, making it almost certain to be a rocky world. What made it truly special was that it was orbiting the nearest star to the Sun. Proxima Centauri is a faint red dwarf star that is just 4.2 light years away. The exoplanet's apparent orbit took 12 days to complete, which put it right in the habitable zone of its parent star.

For two months, Anglada-Escudé and his growing team of collaborators would receive 20 minutes of telescope time per night to track the movement of the star, looking for the wobble that they predicted to be there if the planet truly existed. They also received Paul Butler's agreement to reanalyse the data he had collected on this star.

Anxious to avoid any controversy this time, the team set up a website, www.palereddot.com, that would report their progress in collecting the data but would not publish the data until after the full observing run was over and everything was analysed.

After just 20 days of observation, they could clearly see the planet's signal. It was much clearer that they had dared to hope. That was when Anglada-Escudé suddenly started to feel nervous. 'You get a feeling that this is historic and will be remembered. For a certain few days or weeks you have this unique knowledge that you know will cause a lot of excitement but [your own feeling] is not excitement,' he remembers of that time.

Indeed it did generate great interest. The announcement was made from the European Southern Observatory's Garching headquarters on 24th August and made headlines across the world.[44] It was hardly surprising: the team was announcing a potentially habitable planet around the nearest star to the Earth. Nature could hardly have been kinder. It should even be possible to examine this world directly in the next few years with the JWST space telescope and with the large ground-based telescopes currently being built. In preparation, theoreticians have been busy making predictions about what we might discover about Proxima b, as the planet has come to be known. Unfortunately, not all of those assessments are favourable to the planet being habitable.

One paper[45] looked at the magnetic activity of the star. Proxima Centauri is an extremely magnetic star, about 600 times stronger than the field strength of the Sun, which itself is twice the strength of Earth's magnetic field. In the paper, the researchers from the Harvard Centre for Astrophysics, Massachusetts, found that the magnetic field would create extreme conditions for the planet.

The Sun's magnetic field accelerates a wind of particles off into space. Some of these collide with Earth and cause the aurora to shine in our polar skies. But what's pretty for Earth is positively dangerous for Proxima b. The wind of particles created by its star's vastly stronger field would strike the planet's atmosphere with between one thousand and ten thousand times the pressure of the solar wind at Earth. Although they did not study in detail the effect on the atmosphere, it seems certain this constant bombardment would have stripped the atmosphere away entirely, leaving the planet an airless barren rock. Only by making direct observations as telescopes improve will we know for sure.

Regardless of whether the planet is habitable or not, there seems no question that at some stage in the future, when technology has significantly improved, we will send a space probe there, and in that future century, it will be Anglada-Escudé's name that is spoken.

There was more excitement in February 2017 when Michaël Gillon, of the University of Liège, Belgium, led an international team of astronomers to the discovery of seven rocky planets around a single nearby star. What really excited interest in the collection was that many – if not all – of them could potentially be habitable.

Gillon and his collaborators discovered and studied the collection of planets around the red dwarf star TRAPPIST-1 in two stages. At first they saw just three planets, and this motivated them to conduct an extensive follow-up campaign using a variety of ground-based telescopes and NASA's infrared space telescope, Spitzer. From the data, the team untangled the myriad transit signals they saw, and found that only a seven-planet system could account for it.

Three of these planets orbited in the traditional habitable zone of the star, where liquid water could be present on the surface, but the other four could still be heated enough to have underground oceans, thereby rendering them potentially habitable too.

To underline the powerful world-wide effect of these successes, TIME magazine named Anglada-Escudé, Gillon and NASA's Kepler scientist Nathalie Batalha as three of the 100 most influential people for 2017. They were the only physicists who made the list.

Yet for all the success neither Proxima b nor any of the TRAPPIST-1 planets are the archetypal Earth analogue that astronomers had been hoping for. For that we must hunker down and wait another decade for Plato and its Earth-finding capabilities.

For that we must hunker down and wait another decade for Plato and its Earth-finding capabilities. Or will we? Not if it is down to one of the men who was there at the very beginning, the discoverer of the very first exoplanet 51 Pegasi b.

Didier Queloz's triumphs with the HARPS spectroscope and the onward march of technology mean that he is no longer convinced that the discovery of Earth's twin planet has to come from space.

Back in 1998, Queloz and Francesco Pepe had begun designing HARPS. Five years later, they had it up and running with an accuracy of 1 metre per second for stellar motion. Now it is time to do it all again. This time, Pepe is the principal investigator on a new spectroscope called the Echelle SPectrograph for Rocky Exoplanet and Stable Spectroscopic Observations (ESPRESSO). As the name suggests, the objective for this instrument is to be able to detect rocky exoplanets around nearby stars. To find an Earth analogue would mean achieving an accuracy of between 1 and 3 centimetres per second for stellar motion. This is because Earth pulls the Sun at 10 centimetres per second, so the spectroscope must be able to measure more accurately than this to see an unmistakable signal.

Improving HARPS by a factor of 100 is a daunting challenge but maybe not impossible. In the 12 years that HARPS has been running, Pepe, Queloz and others have learnt more and more about how to use it at its optimum level. They have continued to make little improvements here and there, pushing it to better and better accuracies. Now, on a good night when the night air is crystal clear and the instrument is as finely tuned as possible, they can get readings of 50 centimetres per second. Averaging observations together, which boosts real signals while cancelling out the noise, can get them to 20 centimetres per second.

So Pepe and colleagues have set 10 centimetres per second as their goal for the initial accuracy of ESPRESSO. Scheduled to begin operations at ESO's Very Large Telescope in 2016, Earth-sized planets in the habitable zone of red dwarf stars will be visible from the beginning. This will put ground-based detection capabilities on level pegging with NASA's TESS mission. This is not a bad thing because, where TESS will provide the planets' diameters, ESPRESSO will provide their masses. Astronomers will therefore be able to calculate the planets' densities and know for sure whether they are rocky or gaseous.

The ESPRESSO team will then work to fine-tune the instrument. With HARPS they managed to make it 5 times more accurate. If they do the same with ESPRESSO, they will be at 2 centimetres per second – and that brings Earth analogues truly into range – so long as the stars themselves behave. As the Kepler and CoRoT teams discovered, the unpalatable truth is that now the problem with finding Earth-analogues is the stars.

To be ready in time for ESPRESSO coming online, Queloz is heading an investigation into the stability of stellar surfaces. His team has begun a daily monitoring campaign of the Sun. As this is the only star that can be seen in close-up, the team are hoping to be able to gauge the way changes in the Sun's brightness – slight as they may be – correlate with changes in the movement of its surface. They will then apply what they learn to the Sun's neighbouring stars, when they search for planets.

ESPRESSO will operate on ESO's VLT, which stands in Chile on the mountain top of Cerro Paranal. If or when they find an Earth analogue planet, Queloz jokes that that will be the time for him to retire.

Inevitably, however, the next question to ask will be: is the planet alive? It is not just a new instrument that will be needed to answer this question, but a whole new telescope. A telescope the like of which it is hard to imagine, but which is already being built.

If you stand on the flattened summit of Cerro Paranal and look east, your gaze naturally falls on a range of other peaks. One stands out because its peak is gone, flattened like the one on which you are standing, and the flank is scored with the zig-zag pattern of a dirt track that engineers use to drive jeeps up and down. This is Cerro Armazones where the European Extremely Large Telescope (E-ELT) is being built. Scheduled to open its electronic eyes to the universe in 2024, it is a behemoth. The primary mirror is 39 metres in diameter, more than 4 times larger than the 8.2-metre mirrors in the VLT. Because it is impossible to build a single mirror 39 metres in diameter, E-ELT's will be composed of 798 hexagonal 1.4-metre mirror segments, all held in place by computer-controlled support struts that will constantly adjust the mirrors to hold them in perfect alignment.

Although designed to be a general-purpose observatory for use in all branches of astronomy, the European Southern Observatory admit that it has fully embraced the search for exoplanets.

As a result, not only will E-ELT help detect ever-smaller planets around other stars, but for the larger planets it will be able to take images of some of them and analyse the composition of their atmospheres exactly like the ill-fated space-based observatories Darwin and the Terrestrial Planet Finder were planned to be able to. As with those space missions, in doing the atmospheric analysis, E-ELT has the potential to provide strong clues about whether the planets under scrutiny are living worlds.

Even with a giant telescope on Earth, the challenge of doing this is immense. It will mostly be able to do this work only for the larger, probably gaseous planets, and for nearby rocky worlds such as Proxima b. Only if an Earth analogue is incredibly close will it stand any chance of being analysed in this way.

The James Webb Space Telescope will be doing something similar from orbit following its 2018 launch. Although it will routinely look at super-Earths, it is likely to struggle to analyse true Earth analogues. The only mission that can really do that is a resurrection of the Darwin/TPF idea that NASA is already hoping to position for the 2020 decadal review.

In many ways the discovery of Earth's twin planet will not be the end of the story, but more like the beginning.

Acknowledgements

My thanks to Sam Copeland, Wayne Davies, Kerry Enzor, Richard Green, Richard Milner and Ben Brock; Didier Queloz, Geoff Marcy, Steve Vogt, Malcolm Fridlund, David Charbonneau, Sara Seager, Francesco Pepe, Jon Jenkins; Nicola Clark and anyone I have neglected to mention. Each and everyone of you helped make this book what it is today. I am very grateful for your generous help.

Notes

1 William Hyde Wollaston (1802) 'A method of examining refractive and dispersive powers, by prismatic reflection', *Philosophical Transactions of the Royal Society*, 92: 365–80.

2 Otto Struve (1952) 'Proposal for a project of high-precision stellar radial velocity work', *Observatory*, 72: 199–200.

3 http://www.newscientist.com/article/mg14819992.600-stargazers-amazed-by-crazy-planet.html

4 Joshua Roth (1996) 'Does 51 Pegasi's planet really exist?', http://exoplanets.org/no51pegb.html

5 David Gray (1997) 'Absence of a planetary signature in the spectra of the star 51 Pegasi', *Nature*, 385: 795–6; doi:10.1038/385795a0

6 http://www.nytimes.com/2014/05/13/science/finder-of-new-worlds.html?_r=0

7 http://www.nasa.gov/topics/universe/features/borucki_kepler_prt.htm

8 arXiv:astro-ph/0207133v1 (5 July 2002).

9 Maciej Konacki, Guillermo Torres, Saurabh Jha and Dimitar D. Sasselov (2003) 'An extrasolar planet that transits the disk of its parent star', *Nature*, 421: 507–9.

10 I.W. Roxburgh (2006) 'The quest for a European space mission in stellar seismology and planet finding', 2006ESASP1306..521R

11 Downloaded from: http://www.jpl.nasa.gov/news/fact_sheets/origins.pdf

12 The US Navy had provided Michelson with his education in the first place. He joined the Naval Academy as a midshipman in 1869, and excelled in physics. After graduating, he spent time at sea before becoming a science instructor at the academy in 1875. He resigned from the navy in 1881 to enter civilian academia.

13 A.A. Michelson and F.G. Pease (1921) 'Measurement of the diameter of alpha Orionis with the interferometer', *Astrophysical Journal*, 53: 249–59.

14 http://www.jpl.nasa.gov/releases/2000/sim.html

15 R.N. Bracewell (1978) 'Detecting nonsolar planets by spinning infrared interferometer', *Nature* 274: 780–1; doi:10.1038/274780a0

16 On the same launch was the Planck satellite, which had beaten the STARS exoplanet proposal by just one vote back in 1996.

17 The distance between the Earth and the Sun is 1 astronomical unit (AU) and comes from early 17th-century astronomer Johannes Kepler, who calculated the distance of the planets from the Sun based on fractions and multiples of the Earth's distance. Hence, Venus is at 0.75 AU, Earth is at 1 AU, and Mars is at 1.5 AU.

18 A Planet at 5 AU around 55 Cancri by Geoffrey W. Marcy, R. Paul Butler, Debra A. Fischer, Greg Laughlin, Steven S. Vogt, Gregory W. Henry, and Dimitri Pourbaix, The Astrophysical Journal, 581:1375-1388, 2002 December 20.

19 arXiv:astro-ph/0408585v1 (31 August 2004).

20 Eugenio J. Rivera et al. (2005) 'A ~7.5 Earth mass planet orbiting the nearby star, GJ 876', *Astrophysical Journal*, 634: 625–40.

21 http://www.berkeley.edu/news/media/releases/2005/06/13_planet.shtml

22 'US space scientists rage over axed projects', *Nature*, 439: 768, News 16 February 2006.

23 http://www.berkeley.edu/news/media/releases/2005/09/01_shaw.shtml
24 Dennis Overbuy, 'Finder of new worlds', *New York Times*, 12 May 2014.
25 http://www.nature.com/news/2008/080115/full/451228a.html
26 D. Queloz et al. (2009) 'The CoRoT-7 planetary system: two orbiting super-Earths', *Astronomy and Astrophysics*, 506: 303.
27 http://www.nature.com/news/the-wheels-come-off-kepler-1.13032
28 In case you were wondering, that doesn't mean it is made of styrofoam.
29 http://seagerexoplanets.mit.edu/next40years.htm
30 http://arxiv.org/pdf/0704.3841v1.pdf
31 http://www.space.com/3728-major-discovery-planet-harbor-water-life.html
32 W. von Bloh et al. (2007) 'The habitability of super-Earths in Gliese 581', *Astronomy and Astrophysics*, 476: 1365–71.
33 F. Pepe et al., 'The HARPS search for Earth-like planets in the habitable zone': http://arxiv.org/pdf/1108.3447v3.pdf
34 http://www.nsf.gov/news/news_videos.jsp?cntn_id=117765&media_id=68454&org=NSF
35 Lee Billings, 'The ugly battle over who really discovered the first Earth-like planet' (http://www.wired.com/2014/11/exoplanets/) gives a fuller discussion of this controversy.
36 http://www.nasa.gov/topics/universe/features/rocky_planet.html
37 http://content.usatoday.com/communities/sciencefair/post/2011/01/rocky-exoplanet-reaction-round-up/1#.VZ_hbHjV24t
38 Dirk Schulze-Makuch et al. (2011) 'A two-tiered approach to assessing the habitability of exoplanets' *Astrobiology*, 11(10): 1041–52; doi:10.1089/ast.2010.0592.
39 http://www.nasa.gov/mission_pages/kepler/news/keplerm-20130515.html

40 http://www.skyandtelescope.com/astronomy-news/kepler-goes-down-and-probably-out/

41 Andrew W. Howard et al. (2010) 'The California Planet Survey. I. Four new giant exoplanets', *Astrophysical Journal*, 721:1467–81.

42 Robert A. Wittenmyer et al. (2010) 'On the frequency of Jupiter analogs', arXiv:1011.4720v1.

43 https://www.youtube.com/watch?v=2-WbXT28oQo

44 A terrestrial planet candidate in a temperate orbit around Proxima Centauri by Guillem Anglada-Escudé et al. doi:10.1038/nature19106

45 arXiv:1609.09076v2, 9 Dec 2016.

Glossary

AU Astronomical Unit, the term for the distance between the Sun and Earth: roughly 150 million kilometres.

CCD Charge-coupled device, a detector similar to a digital camera.

CHEOPS CHaracterising ExOPlanets Satellite, ESA mission to discover 'super-Earth'-sized worlds. Due for launch in 2017.

CORALIE Spectrometer built after ELODIE to search for extrasolar planets.

CoRoT Convection, Rotation and planetary Transits, a European mission that was the first space mission dedicated to exoplanet research and designed for this purpose.

ELODIE Spectrometer built by Didier Queloz and Michel Mayor to search for extrasolar planets.

ESA European Space Agency.

ESO European Southern Observatory

ESPRESSO Echelle SPectrograph for Rocky Exoplanet and Stable Spectroscopic Observations, spectrometer to replace HARPS, currently being built by Didier Queloz and Francesco Pepe for use on ESO'S VLT.

ESTEC European Space Research and Technology Centre

FRESIP FRequency of Earth-Sized Inner Planets, a mission proposed to NASA. It was not funded but led to the Kepler proposal.

HARPS High Accuracy Radial-velocity Planetary Search spectrometer, built by Didier Queloz, Michel Mayor and Francesco Pepe to look for exoplanets using ESO's VLT.

HIRES High Resolution Echelle Spectrometer, built by Steven Vogt with exoplanet searches in mind.

IRAS InfraRed Astronomical Satellite, built by NASA, The Netherlands and the UK. Launched in 1983.

JWST James Webb Space Telescope, formerly known as NGST, the successor to NASA's Hubble Space Telescope.

NASA National Aeronautics and Space Administration, the United States of America's national space agency.

NGST Next Generation Space Telescope, since renamed JWST, the successor to NASA's Hubble Space Telescope.

OGLE Optical Gravitational Lensing Experiment, an observational programme to look for celestial objects too small to be seen directly.

PRISMA Probing Rotation and the Interiors of Stars with Microvariability and Activity, a mission proposed to ESA. It was not funded but led to the STARS proposal.

SETI Search for Extraterrestrial Intelligence.

SIM Space Interferometry Mission, a proposed NASA mission to look for exoplanets by precisely measuring the movement of stars.

STARS Seismic Telescope for Astrophysical Research from Space, a mission proposed to ESA. It was not funded but led to the Eddington proposal.

TESS Transiting Exoplanet Survey Satellite, NASA mission to survey the brightest stars for planets.

TPF Terrestrial Planet Finder, NASA's proposed mission to find and analyse the atmospheres of Earth twin planets. Shared the science goals of ESA's Darwin mission.

VLT Very Large Telescope, ESO's observatory at Cerro Paranal, Chile.

Index